TERRAIN FACTORS IN THE RUSSIAN CAMPAIGN

*This publication replaces DA Pam 20–290, July 1951.

For sale by the Superintendent of Documents, U.S. Government Printing Office
Washington, D.C. 20402

Facsimile Edition, 1982, 1986

Center of Military History
United States Army
Washington, D.C.

PREFACE

This study was compiled from a series of reports written especially for the Historical Division, EUCOM, by several former German generals. All of these officers had extensive combat experience during World War II, especially on the eastern front. The principal author, for example, was successively chief of staff of a corps in France and Holland (1939–40), commander of an infantry division in northern and central Russia (1941–43), instructor at a school for division commanders in Berlin (1943), commander of a corps in southern Russia (1943–44), and acting commander of an army in southern Russia (1944).

Terrain Factors in the Russian Campaign describes only the salient geographic features of areas in European Russia actually entered by German troops during World War II, the terrain problems encountered, and German methods of dealing with those problems. Corresponding Russian methods have already been described to a considerable extent in CMH Pub 104–12, *Russian Combat Methods in World War II.* The effect of climate upon terrain is dealt with here only in general terms. Arctic warfare is excluded.

In its original form, this study consisted of a treatise on Russian terrain by the principal author and eleven tactical narratives by other officers, illustrating the effect of various types of terrain upon combat. The description of Russian terrain by the principal author and his observations on the strategic significance of these geographic factors are carefully preserved and find the same expression in the following translation as in the original German. The subsidiary tactical examples have been screened for pertinence to the main subject, for clarity and brevity, and for interest to the American reader. The editors have made every effort to retain the method of presentation, the expressions, and even the prejudices of the principal author. The reader is reminded that all publications in the GERMAN REPORT SERIES were written by Germans and are, therefore, from the German point of view.

CONTENTS

[The first five illustrations are U. S. Army photos from captured German films. The last one is a captured German combat painting now in the custody of the Historical Properties Branch, Office of the Chief of Military History, Special Staff, U. S. Army.]

SECTION I
INTRODUCTION

The influence of rivers, forests, swamps, and steppes on the conduct of military operations is as old as war itself. Ever since wars have been waged, every commander, of high and low echelons alike, has logically utilized natural terrain features to his own advantage and the enemy's disadvantage. The defender, as the weaker party, entrenched himself behind natural lines of resistance which would hamper the enemy's approach and conduct of operations as much as possible. By preference, the defender anchored his exposed flanks on the formidable natural obstacles presented by watercourses or forests. The attacker, on the other hand, sought out terrain favorable for approach marches and combat. He avoided large forest and swamp areas; he outflanked the enemy who sat behind natural or artificial obstacles; he, so to speak, maneuvered the enemy out of position.

A study of military history reveals that prior to World War I very few river crossings were forced in the face of the enemy. The explanation is easy enough. Before World War I armies were relatively small. In most instances, there was no reason for forcing a crossing against any enemy who had chosen a watercourse as his natural line of resistance. The watercourse could be crossed much more rapidly and with less trouble at a place which the enemy had chosen not to hold at all, or at best with only a handful of troops. A similar situation prevailed in the case of large forest and swamp areas. As long as armies were relatively small, exceptional circumstances had to arise before the commander decided to join combat in those terrain obstacles.

With the evolution of armies comprising millions of men, the picture changed overnight. Even the initial stages of military operations are completely different from those of bygone days. An army assembles on a continuous front at its frontier; the enemy can no longer be outflanked and maneuvered out of position. Freedom of action must be gained forcibly by breaking through the enemy front.

In the further course of operations, an army of millions needs space in which to maneuver. Whenever part of the army is confronted by an extensive terrain obstacle, the simple expedient of sweeping around it, as in the days of the smaller army, will usually prove to be impossible because adjacent forces occupy all available space on

---

This publication replaces DA Pam 20–290, July 1951.

both flanks. If natural terrain obstacles in most instances prompted the small army of the past to skirt a particular area, the modern army faces an entirely different problem: The space factor now compels that immense military machine to fight its way through natural obstacles.

For this reason the forced river crossings of World Wars I and II belong, as it were, to the commonplaces of modern military operations. Prior to 1914 no one would have believed that sustained warfare was possible on glaciers, in swamps, or in dense forests. Grim reality proved otherwise. The Argonne Forest and the swampy lowlands of Flanders witnessed 4 years of fighting. The swamps lining the Pripyat area and those along Lake Narocz could no more be bypassed than could the rocks and glaciers of the Dolomite Alps.

World War I showed that modern armies are capable of fighting in any terrain and can hurdle almost any type of natural obstacle during the course of combat operations. One accomplishment alone they found impossible: Neither size nor matériel enabled them to bring the war to a quick conclusion with a succession of rapid blows. The stumbling block was the defensive fire power of modern weapons. Soon after the outbreak of World War I, fire power forced the armies into the ground, and military operations degenerated into a war of position. In the final analysis, trenches are no more than artificial terrain obstacles.

During the interwar period military experts all over the world racked their brains for a way out of that quandary. Some saw the answer in armor, some in the further development of aviation, and others in a combination of armor and aviation. Aviation appeared to be particularly well adapted to the end desired since the airplane is seldom hampered by ground obstacles. Conflicting views on the subject continued up to the outbreak of World War II.

Hitler advocated the solution which later was to be labeled the "blitzkrieg." The idea is very simple and, at the outset, logical: The only way of avoiding a war of position is to stay on the offensive from the first to the last day of hostilities, thereby robbing the enemy of all freedom of action. However, the ability to stay on the offensive throughout a campaign presupposes a massing of forces at few, but well-chosen points. Therefore, there can be no splitting up of armor among the infantry, which must fight its own way, but instead the most thorough concentrations of armored divisions into a few strategic formations which are able to crush every bit of enemy resistance. There must be no dissipation of air power, but a commitment *en masse* of aviation at a few decisive points in close coordination with ground operations, mostly in conjunction with the armored units. Two additional factors become indispensable if the offensive is to be successfully maintained throughout the war: uninterrupted supply and—above all—an unrelenting forward drive by the entire armed forces.

The so-called blitzkrieg brought nothing but success during the first part of the war. The Polish and French Armies, as well as the forces of the Balkan states, were defeated in short order. In no instance did the defending forces succeed in disrupting the continuity of the German offensive. Natural obstacles sometimes had a retarding effect and, on occasion, gave rise to local crises, but on the whole they were surmounted by the momentum of the advance and the skill of command.

It seemed as though natural obstacles no longer played any material role in the conduct of military operations. The great riddle of how to wage war in modern times appeared to have been solved with the blitzkrieg. There can be little doubt that Hitler harbored this same train of thought in embarking on the war against the Soviet Union.

SECTION II

EUROPEAN RUSSIA: A NATURAL FORTRESS

The western ramparts of the Soviet Union, that is to say, the area between central Europe and the Ural Mountains, encompass a tremendous expanse by European standards. Yet, those ramparts comprise only a part of the USSR; its expanse beyond the Urals is even vaster. True enough, western Russia is a geographic subdivision of the Continent, but no other European country can boast the size and number of natural obstacles that protect even such a fraction of the Soviet realm. Any attack from the west must hurdle those very obstacles, and, at the same time, overcome the military resistance of the Soviets. In all that great expanse, only one major river, the Pripyat, flows from west to east and appears to provide access to the interior. But, of all the freaks of nature, just that river and its tributaries form such a maze of swamps that the watershed of the Pripyat constitutes an obstacle rather than a gateway to the interior of the USSR. Practically all other streams and rivers of the Soviet Union flow from north to south, though a few flow in the opposite direction. An attacker approaching from the west thus faces one natural obstacle after the other. As one proceeds toward the east, those obstacles become more and more formidable. The Dniestr, the Bug, the Neman, and the Dvina conform reasonably well to the usual concept of natural obstacles in the form of watercourses, although they are the very rivers that are peculiarly treacherous. The watersheds of the Dnepr, the Don, and the Volga constitute barriers of extreme difficulty. Moreover, the tributary streams of those watersheds combine with the main rivers to form what amounts to a perfect defense system. A look at the tributaries of the Dniestr on a 1:300,000 map, for example, shows that no military architect could have laid them out to any better advantage.

In central and western Europe and on the Balkan Peninsula, not many swamplands have been left in their primitive state. Western Russia, however, still abounds in them. At one place or another, a high embankment may carry one of the strategic railroads through these swamps (in most instances in an unerringly straight line over every obstacle). Some stretches are crossed by a hard-surfaced road, but aside from these few man-made structures the picture remains as nature first painted it. Since the swamplands are part of the river systems, they form a double obstacle. Natural obstacles like the Pripyat area, which blocks the approaches to central European Russia, and swamp and water barriers like those formed by the Volkhov or the lower course of the Volga are not to be found in central and western Europe.

Then there are the Russian forests, most of which merge with the swamplands. Northern European Russia is a woodland interspersed with swamps; the central part of European Russia abounds in forests; the southern part of European Russia is practically devoid of woods. As a matter of fact, European Russia is the only region of the Continent that has arid steppes and sand flats of typical desert character.

Northern European Russia proper, that is to say the swampy woodland north and northeast of the Valdai Hills, is not suitable for mobile warfare, particularly not for large armored formations. The crucial blows of an offensive, therefore, have to fall in central and southern European Russia. In central European Russia lies the Smolensk-Moscow Ridge, a low glacial moraine whose western extension is known to the Germans as the Orsha Corridor (*Landbruecke von Orscha*). This is the watershed between the Black and Caspian Seas in the south, and the Baltic and White Seas in the north. Here are the sources of the Dnepr, the Dvina, the Lovat, and the Volga. Access to this ridge is of paramount importance for any conduct of military operations in the western part of European Russia. But the western approaches to the Orsha Corridor are protected by a wide belt of swamps and forests which extends from the Pripyat Marshes past Velikiye Luki and up to Leningrad. After breaking through this belt, an attacker still faces the watersheds of the Don and the Volga. Even if he has reached the Volga, an enemy coming from the west will find himself only in the outer ramparts of the Soviet domain; before him lie the Ural Mountains, and beyond them, Siberia.

SECTION III
PRE-INVASION PROBLEMS

The Invasion Controversies in Germany

The Germans have very naturally tried to find some explanation of their overwhelming defeat at the hands of the Russians. Quite a few books have been written by former high-ranking German generals and the subject has also been discussed, with considerable bitterness and widely divergent views, in the German press. There are, however, three controversies, not products of hindsight, which were the subject of impassioned debates within the German armed forces both before and after the beginning of the Eastern campaign. These controversies concerned the following questions:

a. Was the war against the USSR necessary?

b. Was the German top command correct in its combat estimate of the Soviet enemy?

c. Was it possible and justifiable to postulate the experiences of the Polish, French, and Balkan campaigns in dealing with the much vaster expanses of the Soviet Union, with their much more numerous and variegated natural obstacles? In other words, do the size and topography of the Soviet Union not preclude continuity of offensive operations from the first to the last day of the campaign—the blitzkrieg?

The question of whether or not the war with the USSR was necessary goes beyond the scope of this study, which deals only with the influence of Russia's topography on the conduct of military operations. The second question has been answered by Hitler's public statement in the Reichstag: He confessed that the magnitude of the Soviet war potential had surprised him. In other words, he had militarily underestimated the Russians. This admission itself contributes part of the answer to the third question: Can the expanses of Russia and the Soviet forces defending them be quickly conquered by means of one sweeping offensive?

Limitations of the Blitzkrieg

The distance from the line of departure of the German offensive on the Bug River to the Orsha Corridor measures about 300 air miles. The distance from the same jump-off line to the middle course of the

Volga measures at least 1,100 miles. Since the attack on the Soviet Union could be launched from only one quarter, namely the west, the Russians were in a position gradually to throw their entire armed might into the contest for the western ramparts of their motherland. The population of the USSR outnumbered that of Germany by more than two to one. While Germany was able to raise an army of roughly 10 million men, part of whom had to be kept in France, Holland, Belgium, Norway, Yugoslavia, Greece, and Africa, the Red armed forces comprised at least 20 million. Thus, the Germans faced not only the vast expanse of European Russia with its numerous and difficult natural obstacles but also an enemy doubly superior in numbers.

A sober consideration of those factors should have revealed the impossibility of overrunning the great area of European Russia and its defenders in one fell swoop. A sustained, uninterrupted offensive had not even been possible in the campaign against France, despite the fact that in that instance neither the area involved nor the enemy's numerical war potential even remotely approached those of the Soviet Union. After the first great battles on the Franco-German frontier the German advance had been forced to halt for about 10 days on the Somme. In dealing with Russia, just a simple and rough estimate of logistical problems made it evident that merely the distance factor, not to speak of Soviet military resistance, would of necessity require protracted breathing spells. Even the single 10-day stop in France had offered the French Army an opportunity for consolidation. There was no reason for assuming that the Red Army would be unable to take advantage of a similar situation. Moreover, should the German forces suffer reverses, a possibility in any war, the very vastness of the Russian land mass held the dire threat that the initiative, which previously had been with the attacker, would be seized by the enemy. Limited as the number of German forces at hand was, they could gain control over the Russian spaces only if they succeeded in continuously carrying the fight to the enemy.

But what if the Soviet command would resort to the venerable and traditional tactics of fighting only a delaying action in the western ramparts of the country and would purposely retire into the depths of the tremendous expanse of the interior? A course of action along those lines, since the Soviets faced an enemy on only one front, might lead to a vicious circle of never-ending hostilities. The situation was reminiscent of that which prompted Bismarck's classic remark to the King of Prussia when, after the victory of Koeniggraetz in 1866, the king wanted to march on Vienna. "Why stop at Vienna?" Bismarck inquired. "Your Majesty might as well go right on to Constantinople and leave Prussia to its own fortunes!"

German Mistakes in Preparing the Invasion

Hitler was sure that he had found the key to a successful attack upon the Soviet Union. His premises of reasoning seem to have been formulated primarily on emotional and political grounds. The fate of the Soviet Union was sealed. The Communist system was doomed, and he, Hitler, was destined by Providence to ring the death knell of that abomination. He was unshakable in his belief that the very first onslaught would topple the whole Soviet structure like a house of cards. After having annihilated the Red Army in gigantic battles at the Soviet frontier, the right wing of the German Army was to reach the lower Dnepr region (industry and raw materials), the center was to occupy the western half of the Smolensk-Moscow Ridge, and the left wing was to reach Leningrad (direct contact with Finland). From this point on, the only remaining task would consist of launching armored and motorized raids into the depth of the USSR in order to assist the native population in crushing the last strongholds of the Communist regime.

Whether or not Hitler's military entourage shared his views on the war with the Soviet Union is immaterial in the final analysis. Views contrary to his went unheeded. Hitler could point to victorious campaigns in Poland, France, Norway, and the Balkans; he had, up to now, always proved to be right, notwithstanding the warnings and apprehensions of the so-called experts, and he posted the successes to his own personal credit. Never before nor afterward did Hitler's prestige and mystic spell attain the heights that they reached in early 1941.

One fact of far-reaching consequence remains to be pointed out. The campaigns in Poland and France were carefully thought out from beginning to end. The campaigns in Norway and on the Balkan Peninsula were improvisations, but both aimed at definite objectives: The crushing of military resistance and the occupation of the entire country. The campaign against the USSR was neither an improvisation nor had its broad strategic outlines been established as in the case of France. Hitler's belief in the collapse of the Communist regime in the interior of the Soviet Union had the effect that the planning stage never progressed to mapping a German advance beyond the previously mentioned regions: The lower course of the Dnepr, the region west of Smolensk, and the Leningrad area. In the actual course of events, the German operations which took up where the planning had left off were improvisations born on the spur of the moment, without being fitted into the pattern of a large-scale plan, and without a clear, or perhaps even a limited objective. Events took precisely that course which had to be avoided in the vast expanses of Russia if the campaign were to be brought to a successful conclusion: The initiative slipped from the hands of the German forces and was seized by the enemy.

It is a safe conjecture that Hitler had made a thorough study of the topography of European Russia. However, in tune with his entire disposition he saw many things not as they actually were but as he wanted them to be. Then, too, he had become spoiled by the successes of the previous campaigns. On several occasions the momentum of an advance had far exceeded his expectations. Natural obstacles seemed to have lost most of their meaning; like difficulties, they were there to be overcome. He simply did not wish to acknowledge the military might of the Soviet Union, even though it was called to his attention by the German military attaché in Moscow and other well-informed persons.

In his ignorance of the Soviet land and people, Hitler was not alone in Germany. This lack of knowledge was common throughout the entire country, even in the top echelons of the armed forces. That particular fact can only be acknowledged, not explained. The Germans have lived near the Russians for centuries. After the partitions of Poland the two nations had a common frontier for 140 years; they have been allies and have fought one another. Nevertheless, that vast country to the east and its people have forever been and remained a sealed book to the Germans. The so-called Iron Curtain had been a fact long before the phrase was coined.

The best example of the lack of knowledge about Russia was the ignorance of the tremendous difficulties resulting from the muddy season. Twice in the course of every year, prior to the onset of winter and again in early spring, the soil of Russia is softened by rain and thaw. The roads become bottomless, and the countryside turns into a morass. In some regions, the boat rather than the *Panje* cart [Ed.: horse-drawn peasant cart] becomes the means of conveyance for travel from village to village. Every Russian peasant is familiar with the situation and prepares himself accordingly. The design of Red Army equipment, in point of ground clearance of military vehicles for example, takes into account these very conditions. Yet the Germans had never learned of that elementary natural phenomenon. Neither the military leaders nor the intelligence agencies were aware of its implications. The field forces were taken completely by surprise by the first muddy season in the late fall of 1941 and encountered, in the fullest sense of the word, bottomless difficulties. Military operations that had been planned or had actually gotten under way became delayed or were foiled altogether. On the highway between Smolensk and Vyazma in late October 1941, for example, 6,000 supply trucks piled up, most of them loaded with ammunition, rations, and fuel for

GERMAN TRUCK marooned after crossing a bridge near Chudovo (Leningrad Front).

the forces advancing on Moscow. Not that the pile-up was caused by a failure to promptly replace demolished bridges with close-by emergency bridges; it was simply a case of the short approach roads to the emergency bridges disappearing time and again into the mud.

Combat units were no less vulnerable to such difficulties than supply units. For example, after the conclusion of the battle for Kiev, Second Panzer Army was concentrated in the Romny-Glukhov-Novgorod Severskij area in preparation for a thrust and break-through in a northeastern direction toward Orel and Bryansk. As the operation progressed, the right boundary of its zone of advance was marked successively by the Seym, Svapa, and Oka Rivers; the left boundary by the Desna, Nerussa, and Navlya. The Desna Basin is heavily wooded and very marshy.

If the attack was to be successful, it was essential that the bulk of the army reach the firm ground and the railway near Orel before the start of the rainy season. A failure to integrate those traffic arteries into the German supply system might have spelled disaster. Second Panzer Army, therefore, decided to take advantage of the short dry-weather period during the last days of September and launch the attack on 30 September 1941—2 days in advance of the general offensive by the rest of Army Group Center. This timing made possible additional support by the Luftwaffe, the bulk of which was to be tied down elsewhere by 2 October.

At first the attack progressed favorably. The 4th Panzer Division captured Karachev on 7 October and Bryansk on 8 October. Then it started to rain. Within a few hours the roads became bottomless. To continue the advance at the previous rate of speed was no longer possible; all movements of wheeled vehicles became extremely difficult. The river valleys and forests were transformed into impassable obstacles which could not be surmounted or bypassed by lateral detours.

The only solution was to gradually tow the army out of the swamps with the help of tracked vehicles and to reach the firm road leading back toward Bryansk via Fatezh-Kromy-Orel.

The only advantage derived from this muddy season of four weeks' duration was that the Russians were equally hampered, despite the fact that with their light equipment, and especially their vehicles, they were better prepared than the Germans to overcome this natural phenomenon they knew so well. Nevertheless, the bulk of the Russian units in the two pockets of Trubchevsk and Bryansk lacked sufficient mobility and strength to escape annihilation. Similarly, the Russian concentration of forces near Yefremov was unable to seriously threaten the exposed right flank of Second Panzer Army.

155-677 O - 86 - 2 : QL 3

GERMAN HALF-TRACK pulling three vehicles uphill on muddy road.

A critique of the German intelligence agencies in their procurement of data on the Soviet Union lies outside the realm of this study. The so-called Handbook on the USSR, which was issued to the field forces shortly before the outbreak of hostilities, probably represented a digest of the information collected by that organization. If we are to believe that this handbook was the authoritative source for estimates of all things Russian, even on the part of the top commanders, then German intelligence did a poor job. The handbook contained little or nothing about the great military significance of the muddy season, which limits conventional warfare in the USSR to very definite periods and thereby influences every phase of the planning and conduct of operations. Instead, it contained blatant examples of false logic and many misleading statements. The Red Army, for example, was supposed to be particularly vulnerable to threats against its flanks and rear. In actual practice, the very first days of combat were to prove exactly the opposite. Other statements went in a similar vein: The Russians are a people with no mechanical aptitude; they are poor organizers; a weak point of the Soviets is their transportation system.

The difficulties lurking in the size and complexity of the Soviet Union, that is to say, the rivers, streams, extensive forests, swamplands, sand, and road conditions, were not omitted from the book. In tune with the times, however, the high command as well as the combat forces tended to underestimate those difficulties. In this connection it might be appropriate to note that terrain intelligence, like any other form of intelligence, requires discriminating and objective evaluation. To report all the facts is not enough; there are too many facts for proper assimiliation unless their relative importance is clearly indicated. Although career officers must make the final decisions, they are not the most suitable people to render such judgments. Even the most able of them suffer from a certain narrowness of professional outlook. The guidance of military historians can save them from dangerous mistakes, and the open-minded attitude of intelligent civilian travelers will sometimes lead to valuable suggestions. The most useful information to reach the troops prior to the initial engagements on the frontier came not from the printed descriptions of the terrain, from agents, or from the largely outdated maps, but from aerial photographs. They were issued relatively late, probably for security reasons.

Training in the Germany Army for combat under Russian conditions of terrain could by no means be termed uniform or systematic. A specialized training program was impossible for many reasons. Plans for the war against the Soviet Union were long kept a closely guarded secret. There were division commanders who, even a week before the beginning of the campaign, regarded their transfer to the

PANJE WAGON: Mainstay of German transportation in the East.

East and the assembly of German forces at the frontier as a precautionary measure or a bluff. Other units were pulled out of the Balkan campaign and shipped directly to the East; they had no time for special training. Then, too, there were no suitable training grounds. Normandy, Champagne, and the Netherlands provide no opportunity for the practice of combat in deep forests, and neither the Jura Mountains nor, for that matter, the whole of central Europe is suitable for training in the unusual aspects of warfare in extensive swamplands and sandy areas. Training in stream-crossing procedure under combat conditions left little to be desired; a high standard of thoroughness in that department was traditional in the Germany Army.

Wherever the training was actually attuned to Russian conditions of terrain, it was instituted on the initiative of individual army, corps, or division commanders. Since only suggestions but no binding regulations came down from higher levels, any kind of special training program bore the personal stamp of the initiating commander. There was no uniformity of procedure. In only one respect did the German top echelon take uniform and systematic action for coping with the special problems of Russian terrain: The combat forces were equipped with light vehicles, and impedimenta were reduced to a minimum. German-issue transportation mostly comprised Polish *Panje* carts drawn by native horses—Russia's conventional means of conveyance. All unnecessary baggage and every bit of the nonessential ballast that soldiers are wont to lug along was left behind at large collecting depots. One measure designed specifically for warfare in the East was the organization of a few so-called light divisions. The infantry and artillery components of these divisions had a table of organization different from that of the standard infantry division. Their equipment consisted in part of pack animals and small, two-wheeled carts. Two battalions of the artillery regiment carried mountain howitzers loaded on pack animals. One disadvantage was that the light division had only two instead of three infantry regiments.

The reduction in the net weight of vehicles affected only horse-drawn vehicles. The design of motorized equipment was not modified to conform to the unique conditions of terrain in the East. The prerequisites for a project of that nature were lacking. German industry was unable to satisfy even a fraction of the demand for cross-country vehicles. It was unable to replace even current losses, as improbable as that may seem in this age of motorization. Of course, Hitler thought a radical change-over in the automotive industry just as unnecessary as a conversion of over-all industry to total war. In his armament policy, as in all other respects, Hitler had committed himself to a short campaign against the Soviet Union.

SECTION IV
OPERATIONS AT RIVER LINES

Every river is a military obstacle that exercises some degree of influence on offensive as well as defensive operations. This influence is particularly pronounced in Russia, where rivers and streams have a number of unusual characteristics. Earlier pages have mentioned that Russia's principal rivers with their numerous tributaries form a succession of natural obstacles to any attack from the west. Forced crossings thus belonged virtually to the daily bread of the German soldier. At the start of the Russian campaign, the German Army was able to commit a body of field forces and an officer corps with peacetime training in the technique of forced river crossings. That picture changed as the campaign exacted its toll of casualties. The raw replacements were a far cry from the thoroughly trained and self-assured fighting men of the peacetime army. In 1941 all major river crossings succeeded, many of them with surprising speed. Later, the crossings no longer went off with the precision of 1941. The reason lay not in the nature of the watercourses, but in the composition of the troops.

While the rivers of Germany and France are more or less similar in nature, each Russian river has its special characteristics. One river provides no yardstick for the next; experiences gained on the Dvina do not automatically apply to the Lovat. One characteristic, though, is common to nearly every Russian stream: The west bank is higher than the east bank. The most notable case in point is the Volga. The advancing German troops thus enjoyed a considerable advantage: Whenever they approached a river, they were on dominating ground.

The Russian rivers, with few exceptions, are not regulated. But lack of regulation is not in itself a hindrance to forced crossings. In many instances it is even an advantage for the attacker. Dense vegetation on the near shore facilitates the approach to the watercourse proper, provides concealment, and permits surprise of the enemy whose visibility and field of fire are limited.

A different situation is presented by a typical feature of Russian terrain; namely, the broad, swampy lowlands that line one or both banks of a river and afford no cover at all. These lowlands are exceedingly common in European Russia. They pose unusually difficult problems for the attacker. A crossing is difficult enough under those circumstances; it exacts a high toll of casualties and consumes a great amount of time. Even more difficult, however, is the construction of bridges and approach roads after a forced crossing. The standard military floating bridge equipage must be freed for the crossing of the next watercourse and a semipermanent bridge constructed. Moreover, a bridge that merely spans the stream itself is rarely sufficient. All or part of the swampy lowland must likewise be bridged by structures high enough to clear the spring floods and strong enough to resist battering by cakes of floating ice. Since construction on such a scale involves enormous expenditures of labor, the German forces tried their best to seize permanent bridges in flying column actions and *coups de main.*

Many such attempts met with success. For example, at the end of August 1941, Second Panzer Group received orders to thrust southward from the Roslavl-Smolensk area. Two panzer corps, in conjunction with Second Army advancing on the right, were to encircle and annihilate the Russian army group in the Kiev area. The first objective was to be Nizhin, the next one, Romny. This meant that the Desna and the Seym, as well as a number of smaller rivercourses, had to be crossed. The Desna flows through a primeval river valley which is covered by meadows with marshy subsoil. The river itself presents no insurmountable obstacle. But every rainfall turns the banks into such a quagmire that a crossing would have become impossible unless a bridge and its long approach roads could be captured intact. A wooden bridge spanned the Desna near Novgorod Severskij, south of Bryansk. The actual bed of the Desna at that point is only 100 yards wide, yet the bridge was over 600 yards long. Several tanks of Second Panzer Group not only succeeded in getting across that bridge in broad daylight but also caught the Russian antitank gunners on the far shore completely off guard. The bridge, the only one for miles around, fell undamaged into German hands. Had the bridge been blown and a forced crossing, as well as the construction of a new bridge, become necessary, it is doubtful whether Second Panzer Group could have closed the east side of the Kiev pocket in time. As things turned out, half a million Russian prisoners were taken in that pocket.

The major rivers, of course, were the objects of primary attention for the German higher command. Even the map showed them to be the best-defined phase lines for the advance and, at the same time, the most obvious obstacles. In 1941 those rivers were frequently crossed with surprising ease. The wide lower course of the Dnepr, for example, promised serious difficulties. Aerial photographs showed a formidable stream, dotted with islands, and flowing through a wide valley. In actual practice the crossing was a speedy operation. It turned out that the reservoir near Dnepropetrovsk makes the Dnepr

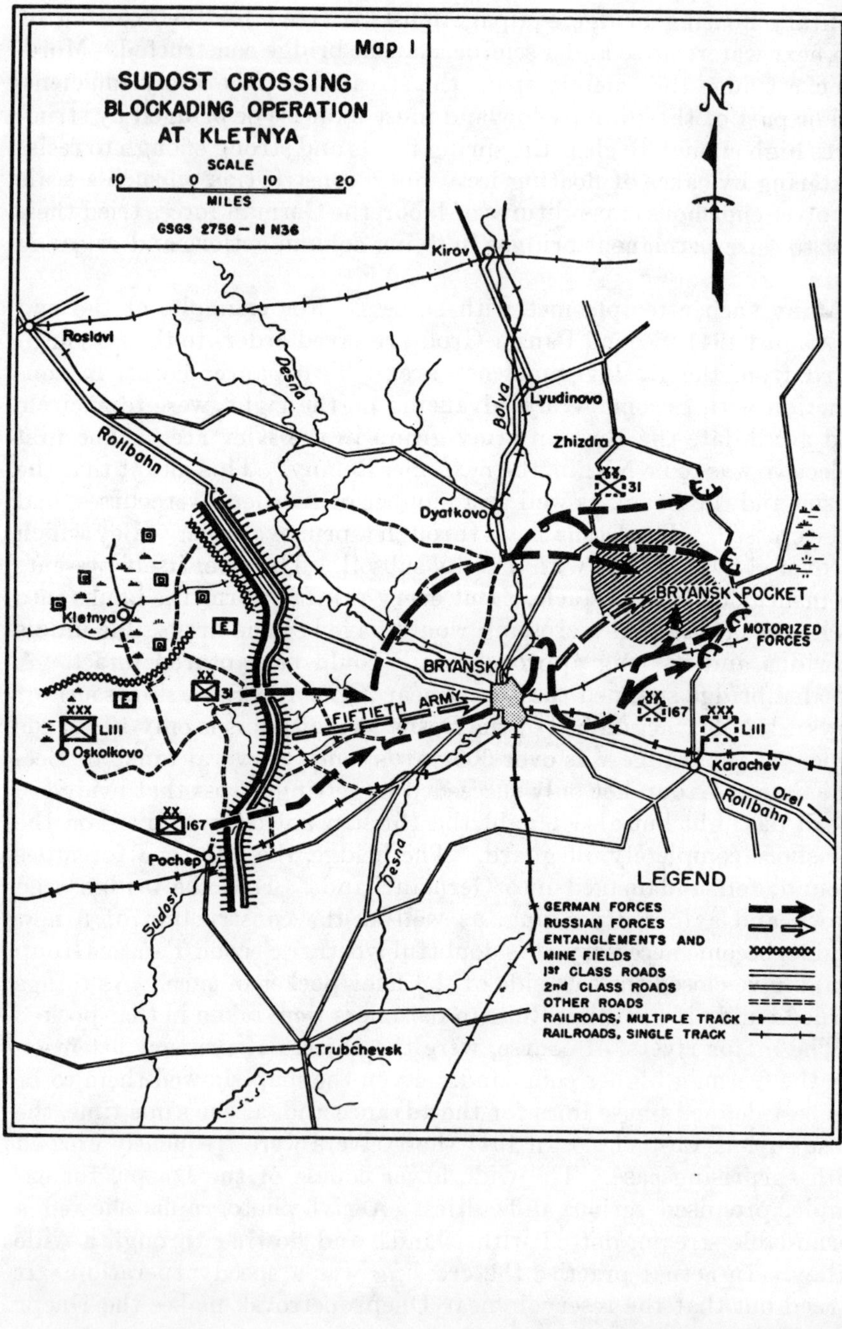
Map I
SUDOST CROSSING
BLOCKADING OPERATION
AT KLETNYA
SCALE
10 0 10 20
MILES
GSGS 2758 — N N36
Kirov
Rosldvl
Desna
Bolva
Lyudinovo
Zhizdra
Rollbahn
Dyatkovo
31
Kletnya
BRYANSK POCKET
MOTORIZED FORCES
BRYANSK
FIFTIETH ARMY
XXX
LIII
167
Oskolkovo
Karachev
Orel
Pochep
Sudost
LEGEND
GERMAN FORCES
RUSSIAN FORCES
ENTANGLEMENTS AND MINE FIELDS
1st CLASS ROADS
2nd CLASS ROADS
OTHER ROADS
RAILROADS, MULTIPLE TRACK
RAILROADS, SINGLE TRACK
Trubchevsk

relatively shallow and creates several ford-like passageways. The Dvina, on the other hand, which looked like a much less serious obstacle, presented far greater difficulties. Finally, a river such as the Volkhov, which is nothing more than a flowing swamp amidst a marshy wilderness of woods and brush, presented almost insurmountable problems. The difficulties were not solely due to enemy resistance, but also to the fact that the velocity of current and the nature of the banks differed greatly from one place to another. The problem was greatly simplified whenever, as in the following example, a period of static warfare allowed time for a thorough terrain reconnaissance.

During the major offensive that began on 2 October 1941, a German infantry corps was given the mission of breaking through the belt of field fortifications in front of Bryansk from the southwest and capturing the city proper. The first phase of this operation was the crossing of the Sudost River, which for several weeks had been the last natural obstacle separating the German forces from those of the Russians (map 1). All necessary terrain intelligence about the river line had been obtained. Patrols of the various arms and services had done a thorough job of reconnaissance. As a result, it was known that the river was not very wide; that it had a firm bed and solid banks, partly formed of clay; that several fords existed; and that a number of cliffs on the far shore were fortified against tank attacks. In addition, deep tank traps had been observed. According to statements obtained from prisoners, the enemy fortifications consisted mainly of well-covered earth bunkers, in some places even of concrete pillboxes, and extended to a depth of several miles. Reconnaissance by captive balloons and by aircraft served to complete the picture.

Although a relatively unimportant river, the Sudost barred access to an enemy system of field fortifications. For several weeks the two divisions of the corps had been in position along the river. The mission of the corps was, therefore, primarily a matter of effecting a break-through on a wide front in an enemy fortification system, the small river in front of it merely representing the first obstacle. Both divisions were to cross the river simultaneously in many spearheads. Regimental and battalion assault teams were formed and liberally provided with stream-crossing and emergency bridging equipment. Engineer troops were assigned to them to facilitate the crossings and carry out demolitions. It was hoped that some of these teams would quickly succeed in gaining a foothold on the far bank of the river and then immediately push on into the fortified enemy lines. The attack was to be launched from fixed positions, and no special deployment was necessary. Auxiliary crossing equipment, primarily foot bridges floated on pneumatic boats, could be prepared almost to measure and was made ready directly behind the crossing points. Other equipment included log rafts for heavy weapons and wooden bridge sections designed to get motor vehicles across narrow parts of the river and over tank traps on the far shore. The Germans dispensed with artillery preparation directly before the attack. Instead, a fire for destruction was initiated well in advance against enemy installations on the other side of the river. The divisions were ordered to launch a surprise attack at dawn, cross the river in several spearheads, and then continue through the enemy zone of fortifications in a deep thrust toward Bryansk. Since there was no contact with other German forces to the right or to the left, the task could be carried out with complete freedom of action. What actually mattered in this instance was that at least one of the spearheads should be able to advance steadily.

The very first thrust carried the troops across the Sudost River and past the enemy antitank ditches. The enemy bunkers, however, were stubbornly defended. Some of them had been bypassed by the assault teams and held out with great tenacity for another 2 to 3 days. For this reason little ground was gained during the first 3 days. On 5 October a Russian tank brigade comprising about 40 tanks attempted a counterattack against the division on the left. Stukas and antiaircraft artillery quickly moved in and successfully opposed the attack. The thorough preparations which had been made for crossing the Sudost proved worth while. The river offered no major difficulties and caused no particular delay in operations. Neither strategic plans nor tactical developments were in any way influenced by it, nor did it have any effect on the performance of the various combat arms.

The air force was not brought into play until after the breakthrough of the enemy system of fortified positions was accomplished. It was plainly demonstrated that the most significant obstacles during and immediately after any river crossing are the enemy's will to fight and the use which he makes of the river as an integral part of his plan of battle.

In Russia there was no way of gaging the prospective difficulties of a forced crossing by the size of the river alone. Insignificant branches of tributaries, mere incidental features on the map, were frequently the very obstacles that brought the advance to a standstill. At one place a half-day's summer rain may ineffectually soak into the ground; but at another it may be enough to transform the stream valley into a spongy swamp and coat the roads with a slippery, soaplike surface. Under such conditions, motor vehicles will slide wildly down into a valley if they do not overturn. But they will be absolutely unable to negotiate the upward grade on the far side. The result is all too familiar: an insignificant, small, and moderately steep slope which defies the most skillful driver; marooned prime movers; long lines of mired vehicles in the valley and on the approach roads. In many instances only the sun can remedy such a situation. Sunshine restores the slope and roads to usable condition just as quickly as the brief rainfall had made them impassable.

There simply are no pat formulas for resolving the many unknown quantities in the equation that is Russia. Surprises are on each day's bill of fare. In the end, the German Army of 1941 always managed to get the situation under control. Not only did every unit have its share of well-trained men, but, above all, these men were led by energetic and self-reliant officers and noncommissioned officers. The following example illustrates several of these unexpected factors.

Late in July 1941, LIII Infantry Corps was ordered to cross the Dnepr between Zhlobin and Rogachev (p. 22). Specific information about the river, even its width, was lacking. Equally sparse was information about the Drut, a smaller river west of the Dnepr near Rogachev. Ground observation was impossible, since the enemy still held the last five miles west of the river. The corps headquarters, therefore, had to obtain the necessary information in the course of the action. Air reconnaissance and the interrogation of prisoners and local habitants were conducted as usual. Liaison with other units, especially the engineer railway troops, provided much useful information.

By D-day it was known that the Dnepr had a swift current; that the river bed was partly swampy; and that, in case of rain, there were impassable stretches along the banks. The width of the river was assumed to be between 150 and 200 yards; an unfortunate error because the Dnepr turned out to be considerably wider in the Zhlobin area. As a result, the bridge equipage provided for in the original plan proved to be insufficient. Partly submerged floating bridges across the Drut and the Dnepr had not been spotted in advance. They were to prove very valuable later on. Prisoners had mentioned their existence; but since the German forces were unfamiliar with bridges of that type, they had failed to pay any attention to the reports.

Strong enemy forces, entrenched in hasty field fortifications 5 miles west of the Dnepr, had to be thrown back before the corps could cross the river. The main attacks of the three divisions—the 267th, 255th, and 55th—were to hit enemy sectors that were directly in line with the prospective crossing sites. It was hoped that the momentum of the attack would carry the divisions across the river, close on the heels of the fleeing enemy. Considerable resistance was to be expected. There could be no question of a surprise attack.

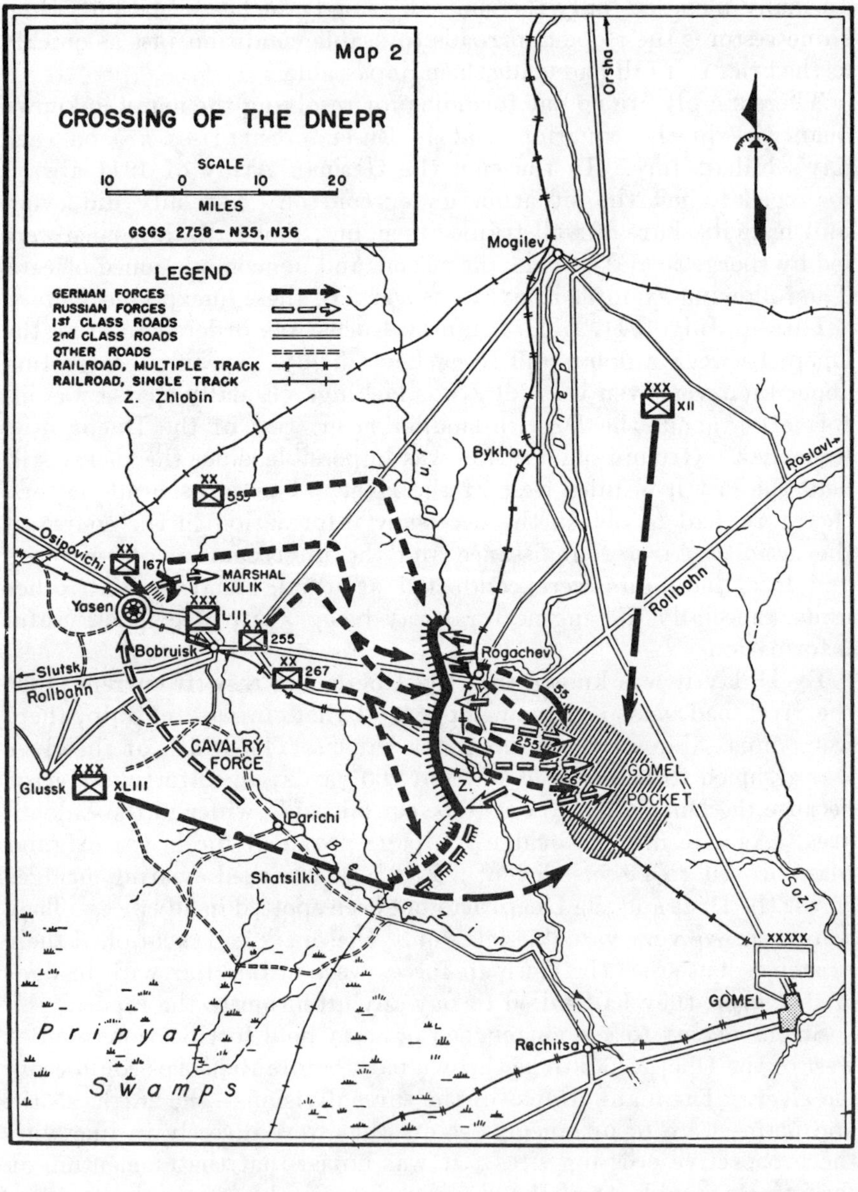
Map 2
CROSSING OF THE DNEPR
SCALE
10 0 10 20
MILES
GSGS 2758 – N35, N36
LEGEND
GERMAN FORCES
RUSSIAN FORCES
1st CLASS ROADS
2nd CLASS ROADS
OTHER ROADS
RAILROAD, MULTIPLE TRACK
RAILROAD, SINGLE TRACK
Z. Zhlobin
Orsha
Mogilev
Dnepr
Bykhov
Drut
Roslavl
Rollbahn
Osipovichi
MARSHAL KULIK
Yasen
Bobruisk
Slutsk
Rollbahn
Rogachev
CAVALRY FORCE
GOMEL POCKET
Glussk
Parichi
Shatsilki
Berezina
Sozh
GOMEL
Rachitsa
Pripyat
Swamps

The German forces did not succeed in crossing the river while the enemy was being thrown back. Therefore, a planned river crossing had to be carried out on the following day. The three divisions of the corps had reached the Dnepr at the selected crossing sites in the course of the attack. Strong spearheads had taken the objectives along the river line and were located at the following points: at Zhlobin, the 467th Infantry Regiment of the 267th Division; in the center, the 465th Infantry Regiment and behind it the 475th Infantry Regiment of the 255th Division; and at Rogachev, two infantry regiments of the 55th Division. This disposition of forces was not changed for the crossing. All three divisions committed the bulk of their artillery in close support of the assault forces. Since earthworks had been spotted on the far shore of the river, especially at Zhlobin, strong air support by Stukas was arranged for. Increased air reconnaissance became necessary since any ground reconnaissance was out of the question as long as the enemy was still holding the near shore of the river. Preparations in the vicinity of the river were equally impossible, and for this reason pneumatic boats and auxiliary equipment were loaded on trucks and kept in readiness to be brought forward quickly to the points where the troops would reach the river Because of the discovery that the available bridging material was insufficient for the construction of a military bridge across the Dnepr at Zhlobin, an additional bridge train was requested from Army and arrived in time. This was a fortunate decision; precisely the length of the extra bridge train would have been lacking otherwise.

The two divisions on the flanks—the 267th Division at Zhlobin and the 55th Division at Rogachev—each concentrated their attacks on a single objective. The 255th Division, rather widely spread out in the center, formed two points of main effort, one toward the east and the other toward the southeast. To free troops for the attack, extensive use had been made of mines during the last few days. At one point, in the case of the 55th Division, an entire infantry regiment was detached from the front, leaving only one company on line to deceive the enemy.

By afternoon of 14 August 1941, Zhlobin had been captured by the 467th Infantry Regiment of the 267th Division, the western bank of the Dnepr had been reached by the 465th Infantry Regiment of the 255th Division, and the 55th Division had crossed the Drut River and captured Rogachev. At that point the various moves preliminary to the Dnepr crossing were completed. Very heavy fighting still raged around Zhlobin and Stukas repeatedly had to go into action against the railroad embankment west of that town, which commanded the entire area.

On 15 August 1941, as XII Infantry Corps east of the Dnepr maintained steady pressure on the enemy to the south, the 55th Division succeeded in crossing the Dnepr ahead of schedule since enemy resistance was weak at this point. However the enemy defended himself that much more stubbornly north and east of Zhlobin. Not until the afternoon of 15 August, and only after repeated Stuka attacks, was it possible in that area to gain a foothold on the eastern bank of the river.

The Dnepr presented a natural obstacle in the path of the advance only in the southern sector of LIII Infantry Corps, and there only to a certain extent. By the time the corps had reached the Dnepr, the push of XII Infantry Corps in the northern part of the zone of advance had become so effective that the enemy offered no further decisive resistance at that point. This experience demonstrated anew that the effect of rivers depends largely upon the enemy's dispositions and his determination to fight.

On the southern sector the enemy at first offered stubborn resistance. The swiftly flowing river, 150 to 200 yards wide, proved a difficult obstacle. An insufficient amount of equipment was available for the crossing, and even the effect of massed artillery did not shake the enemy. Stuka attacks had to be made repeatedly. The retarding effect of the Dnepr was enhanced by the enemy's will to resist. But if the struggle for the river was unduly prolonged, the German troops were partly to blame. Even after reconnaissance patrols of the engineer railway forces had gone across at Zhlobin, the infantry along the river was still trying to decide whether this was the right time to make a crossing. This hesitancy brought about a delay of half a day and made it possible for parts of the enemy forces to be withdrawn from the Gomel pocket.

The longer the war lasted and the higher the toll of casualties mounted, the smaller became the proportion of well-trained, professional soldiers in the German fighting machine. This fact cannot be ignored in looking at the later stages of the campaign when the tide turned and the enemy seized the initiative. In the defensive actions and withdrawals to follow, the numerous successive river lines could have become the backbone of the German defense. However, they were not utilized to that advantage because the troops had become numerically too weak. Here was an object lesson of the dire consequences threatening any army that invades a large country in which it must fight successive battles with a numerically superior enemy. The very moment that the offensive grinds to a halt and cannot be resumed full-scale, the previous attacker is compelled to follow the dictates of the enemy. In a narrowly delimited area there always remains the possibility that a certain equilibrium can be maintained in the give-and-take of battle; in a region as tremendous as European

Russia, however, such a turn of events may result in a catastrophe for the numerically inferior army.

Previously, it had been the German forces and the German command who had carried the fight to the enemy, bypassed forests, and skirted swamplands. Now the Russians were doing it, but with far superior forces. And when air reconnaissance fails in such a situation, the defender gropes in the dark. He has no way of knowing from which side the enemy is coming. He should be strong everywhere, so as to be ready for any eventuality. But where was the German Army to get the necessary manpower? In order to maintain any kind of an unbroken line over a long defensive front, individual units were forced to hold wider and wider sectors. By 1944 a frontage of 30 miles for an infantry division no longer caused even a raised eyebrow. In terms of men per yard of frontage, this amounted to one defender for every 30 yards. A river sector with a defense no more adequate than that constitutes hardly more than a water obstacle for a strong enemy. This will explain why, during the German withdrawals in the last part of the war, the watersheds of the Russian rivers no longer played the role for which nature had typed them and which the German top command had expected them to play.

The German Army now had to pay for the fact that not one of the Russian river lines had been turned into a fortified rearward wall. The Dnepr and, as a northern continuation, the Dvina or Lake Peypus, fairly begged utilization to that advantage. During the course of the war the field forces time and again clamored for the establishment of such a fortified zone. The underlying idea was not to set up a mere defensive river line, but rather to create a rearward buttress that simultaneously would serve as an advance base for all military operations in the East. That no steps were ever taken toward establishing such a line was allegedly due to the shortage of labor and materials. The establishment of temporary rearward positions behind river lines was likewise a little-used expedient in the Russian campaign. The tremendous distances caused manpower shortages everywhere, above all at the front. Hasty field fortifications of the conventional type, like machine-gun and artillery emplacements or foxholes, are little more than useless in Russia. If a position is to be held through the severe Russian winter, heated personnel shelters are indispensable. But the construction and maintenance of a large number of personnel shelters requires a great amount of manpower, and manpower was invariably in short supply.

During winter the watercourse loses its value as a natural obstacle because it freezes over. The Russian streams are generally icebound from December into April and will even bear the weight of tanks. Many experiments were conducted toward neutralizing that disad-

TYPICAL RUSSIAN WOODEN BRIDGE *across the Drut River.*

vantage. Attempts were made to let wire obstacles freeze into the ice in the middle of the river or near one of the banks. A single snowfall or blizzard, however, is apt to nullify the effect of such an obstacle, not to speak of the fact that defense works of that nature are feasible only in rearward positions and can never be set up in the face of hostile forces occupying the far shore. All mines except those equipped with extension trip rods (*Stabminen*) are neutralized by deep snow. Demolition charges in a variety of shapes were used for blasting at least narrow channels into icebound streams. If a Russian cross-river attack was known to be imminent, German artillery tried training its fire on certain points of the ice cover, so that the resulting ribbons of open water might help to canalize the enemy forces into predetermined sectors. In the final analysis, however, all those were fruitless efforts. The cold was always the victor in the unequal struggle. There was no getting around the fact that nothing can be done about the ice, and that a river, once it is frozen, loses all value as a natural obstacle.

In the spring of 1942, when warm weather began to thaw the unusually deep snow cover, the German Army faced yet another surprise. Since the winter had been very severe, the rivers carried heavy ice floes which simply crushed the newly built emergency bridges. That surprise could have been guarded against by a mere look at whatever Russian wooden bridges were still intact. As flimsy as their superstructure might be, their substructure was invariably solid and protected by strong, sharp-nosed ice fenders. Well-anchored underwater bridges also rendered excellent service even under flood conditions. They were widely used by the Russian Army.

Fording is impossible during floods, except in the case of small creeks; river crossings by boat are very dangerous. Even skilled boat crews were sometimes lost. Mechanically the German assault boat with shallow-draft outboard motor gave good service in flood waters; other types of marine engines were clogged by the sand and sediment in the water. Such difficulties were of particular importance for the establishment and retention of bridgeheads. Failure to take the necessary precautions entailed the risk of having bridgeheads completely cut off from all supplies. For this reason, both Germans and Russians repeatedly evacuated small bridgeheads before the onset of floods.

SECTION V

OPERATIONS IN WOODS AND SWAMPS

World War I clearly proved the German need for training in forest fighting. Early in that struggle the French *Chasseurs Alpins* demonstrated their superiority over German infantry in the close terrain of the Vosges forests. But the period between wars saw that technique of combat treated as the same stepchild it has always been. The authoritative German field manual on operations bears witness to that fact. Whereas the tactical principles governing combat in open terrain are clearly and unmistakably worked out, the sections dealing with the technique of combat in woods have somehow a flavor of mere theory. The authors of the manual simply lacked the necessary flair for forest fighting.

This reason, if no other, helps to explain why the German higher and lower command had a tendency toward avoiding forests and forest fighting during the Russian campaign. But since northern European Russia—the region north of the Valdai Hills—is one vast woodland interspersed with numerous swamps and since central European Russia is likewise heavily wooded, the war could not be waged without much forest fighting.

Large, continuous forests that are frequently and extensively interspersed with swamps are characteristic of Russia in general. Those woodlands actually begin as far west as Poland. The large forests of Augustow and the Carpathians are remembered for their role in World War I. The forest of Byalovizh, formerly a hunting preserve of the Czars and later of the Polish Government, differs from the other woodlands to the extent that, because of its special use, it has been made accessible in all parts by an adequate network of roads and paths. Farther to the east there follows the large and deep forest and swamp zone of the Pripyat region. Prominent among the contiguous, heavily wooded areas are the wide belts around Minsk, Borisov, Orsha, Vyazma, Bryansk, and Gomel, and north of the Smolensk-Moscow Ridge, there are forests surrounding Polotsk and Velikiye Luki. The Ukraine, which was still steppeland during Mongol times, and the adjacent part of southern Russia are, except for the spurs and the subalpine ranges of the western Caucasus, distinguished by their singular lack of woods.

No universally applicable description can be written of these forests which usually merge with stretches of marshland. Soil and terrain differ from place to place and are subject to seasonal and climatic influences. Maps often give a distorted picture. The Pripyat area, for example, not only contains patches of sandy soil with tall stands of timber, but also well-populated stretches of cultivated land with country roads built on dikes. Surrounding those islands of civilization are all types of marshes and swamps: bare, mile-long reed flats and weed-covered bogs which are completely impassable even on foot; wet birch and alder swamps through which man can barely wade; treacherous, meadow-like flats, whose short, dark turf yields to the slightest pressure and swallows a motor vehicle to its very top. Whatever the map shows to be marshland may or may not become perfectly passable after a long spell of hot weather. A sudden summer storm can wreck all plans. The countless large and small watercourses require an infinite number of bridges.

In the summer of 1941 many parts of the large forests of Minsk, Borisov, and the Smolensk-Moscow Ridge were completely dry. During weeks of marching through those forests, the German armor and infantry columns were enveloped in clouds of dust. Water for men and horses could be obtained only with difficulty. Each of the few improved roads was designated a *Rollbahn* [main axis of motor transportation from which all animal transport and marching columns were normally barred] and as such was reserved for armored and motorized units. The inevitable result was that certain stretches of those traffic arteries were soon completely plowed up. The infantry divisions, which had to depend on forest trails during part of their advance, were brought to the verge of exhaustion by marches through the deep, powdery sand. The large forests around Vyazma and Bryansk are nearly impenetrable during summertime because of their dearth of roads and their stretches of swampy soil.

The characteristics of the forests around Lake Ilmen, on the Crimea, and in the Caucasus are entirely different from those of the previously mentioned woodlands. Lake Ilmen feeds and is fed by streams which, except for the Msta, all flow from south to north. Some of the principal watercourses of this area are the Lovat, the Pola, and the Volkhov. Villages are to be found along the rivers, because these are the primary sources of water supply. Since wells must be drilled through a layer of clay and have to be sunk very deep, in some places more than 150 feet, they are scarce in that part of the country. Wet woodlands interspersed with swamps cover the areas between the rivers. Short and tall stands of timber cast their reflections in stagnant pools of ground water. Roads, most of them worse than poor, generally follow the river lines. East-west roads through the woodlands are few and far between. After military operations north and south of Lake Ilmen degenerated into a war of position and after a German thrust in the spring of 1942 during the high point of the thaw had blasted a connecting corridor to the forces caught in the Demyansk pocket, the construction of corduroy roads became imperative. The most widely known was the so-called Reich Corduroy Road which led from southeast of Staraya Russa through the connecting corridor (at its narrowest point only 2 miles wide) into the pocket of Demyansk. Countless hours of toil went into the construction and maintenance of the arterial corduroy roads and their extension up to the very front-line companies. The job was all the more difficult because the uninterrupted defensive actions had reduced the companies to a strength of 30 or 40 men.

The Yaila Mountains in the Crimea bear fine, tall stands of timber, and some woodlands in the heart of the Caucasus region are equally beautiful; but parts of the forests in the Caucasus foothills, particularly in the vicinity of the coast, are impenetrable thickets that grow to twice or three times the height of a man. The crucial drive on Tuapse during the German Caucasus offensive failed not only for want of adequate forces but also because of the difficult terrain. The area is peculiar in that, although the rivers have cut deep valleys, this erosive action has brought no rocks to the surface anywhere. Consequently, in wet weather the few roads become chutes of slippery clay wherever there is a gradient and hopeless bogs where they pass through river bottoms.

Vast, uninterrupted swamplands are relatively scarce in southern Russia. They may be found in the partially wooded lowlands of the Dnepr at Nikopol and Kakovka and in the treeless lagoons at the mouths of the Kuban, Don, Terek, and Volga Rivers. The lower Volga region is entirely covered with marshes.

Central and northern European Russia abound with swamps which, in this instance, merge with extensive woodlands. The forests and marshes of the Pripyat watershed rise like a bastion to protect the gateway to central European Russia. An attacker approaching from the west must of necessity split his forces into two parts. The offshoots of the Pripyat region reach as far as the east bank of the Dnepr. Toward the southeast they extend into the Kiev area, and toward the east they stretch to Gomel. Two strips of marshland adjoin the Pripyat region in the north: a western strip from Molodeczno (northwest of Minsk) through Polotsk to Lake Peypus and an eastern strip from Velizh through Velikiye Luki to Lake Ilmen and Lake Ladoga. The eastern strip of marshland, over which German forces never gained full control, played an important and fatal role in the campaign against the USSR. From the very outset it separated Army Group

Center from Army Group North. This great gap was never completely closed. The swamps prevented any coordinated effort by the two army groups, and only because of that circumstance were the Soviets able to launch their later operations at Rzhev and Vitebsk against the left-wing of Army Group Center. In the final analysis the whole of Army Group North bogged down in that belt of marshes and forests which guards the approaches to the strategic Valdai Hills. Today it is useless to ponder what the result might have been if that belt had been successfully pierced. The mere existence of this great swampland was a decisive factor in the late war.

During the German advance of 1941, the German higher and lower commands revealed a tendency toward bypassing the large forests and skirting around the swamps. In a country as rich in forests and swamps as western European Russia, however, such tactics are possible only to a limited extent. The lack of open space soon forced the troops to engage in forest fighting. The spearheading armored divisions hugged the improved roads in breaking through the forests and sought battle in open terrain. The result was that, having been outflanked and his lines having been pierced, the enemy retired laterally into the depth of the woods. While the immediate vicinity of trails and improved roads had thus been cleared of hostile forces, the German armor had no sooner surged past than the Russians once more emerged from the forests and reformed their lines. The infantry divisions, which trailed a considerable distance behind the tanks, subsequently came up against the same, at times quite sizable, Soviet forces. In the haste of their advance the infantry divisions were likewise unable to do a thorough job of mopping up the enemy in the extensive woodlands. Not that an attempt of that sort promised any success; the cut-off Russians never thought of laying down their arms. They merely fought rear guard actions and withdrew deeper and deeper into the woods. These scattered enemy elements and the large quantities of equipment that had been left in the forests were soon to become the backbone of the partisan units. Events took a similar turn in sectors in which no armored units had been committed. The infantry spearheads were likewise unable to clear the large forests completely of the enemy. After their lines had been pierced, the Soviets simply retired into the depths of the forests and swamps and continued to fight. The following examples illustrate two different methods of dealing with such a situation.

LIII Infantry Corps entered large forest areas at the very beginning of the Russian campaign. Within a few days two different divisions of the corps received precisely similar missions—the protection of main highways against flank attacks launched from deep forests. The first attack at Maloryta was liquidated with great difficulty and

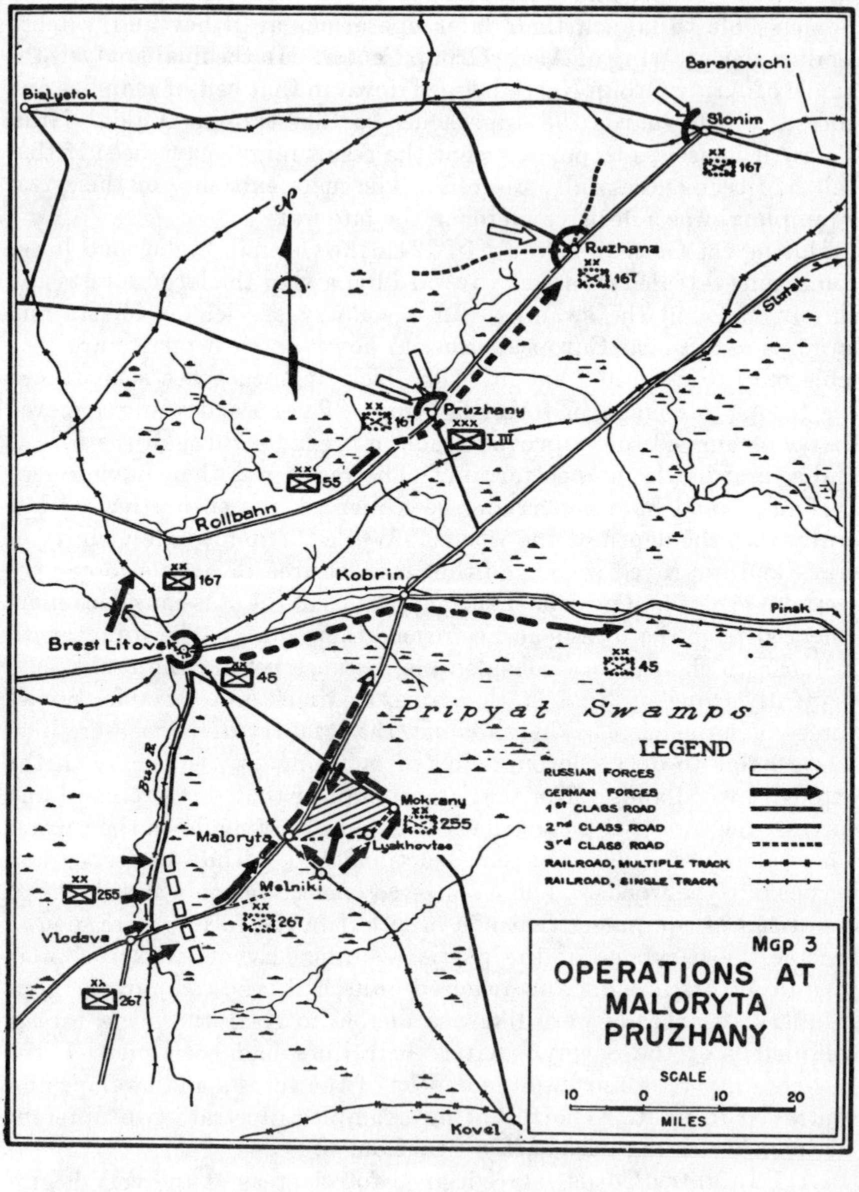
Baranovichi
Bialystok
Slonim
167
Ruzhana
167
Slutsk
Pruzhany
167
LIII
55
Rollbahn
167
Kobrin
Pinsk
Brest Litovsk
45
45
Pripyat Swamps
Bug R.
Mokrany
255
Maloryta
Lyakhovtse
Melniki
255
267
Vlodava
267
Kovel
LEGEND
RUSSIAN FORCES
GERMAN FORCES
1st CLASS ROAD
2nd CLASS ROAD
3rd CLASS ROAD
RAILROAD, MULTIPLE TRACK
RAILROAD, SINGLE TRACK
Map 3
OPERATIONS AT MALORYTA PRUZHANY
SCALE
10 0 10 20
MILES

many German casualties. Four days later, an attack at Pruzhany was blocked and frustrated by the use of entirely different tactics. (See p. 34.)

At Maloryta the mission of the 255th Division (later, because of enemy opposition, reinforced by the 267th Division) was to open up the Vlodava-Maloryta-Kobrin highway and to protect the right flank of Fourth Army (map 3). Because the area of Maloryta was marked on the map as part of the Pripyat Swamp, the 255th Division had expected to find a very swampy forest. Actually, this is an area of old, uncultivated, high forest, with much underbrush and no swampy characteristics. The operation developed out of a meeting engagement with an enemy division that was advancing on Maloryta from the southeast. No special measures of any kind for reconnaissance or security had been taken. A prearranged deployment was therefore impossible and German units went into action wherever they happened to encounter the enemy.

On 23 June 1941, the reconnaissance battalion of the 255th Division was sent out toward Mokrany. At Melniki the battalion came upon enemy scout tanks, which were driven off. During the evening hours the 465th Infantry Regiment occupied the entire forest of Melniki. Meanwhile the 475th Infantry Regiment was stopped outside Maloryta. On 24 June near Lyakhovtse, various enemy attempts to break out toward the east and southeast were broken up, mainly by artillery, which had come up in the meantime and now fired from the route of march. By that time the 465th Infantry Regiment had reached the scene. Since no progress was being made at Maloryta, the commander of LIII Infantry Corps went personally to Melniki and ordered an immediate attack from the east and southeast on Maloryta to be executed by the 465th Infantry Regiment. Elements of the 455th were to take part in this action. He then continued on to the 475th Infantry Regiment which was preparing to launch another attack on Maloryta from the southwest. This attack encountered little resistance and was continued through the town up to the crossroads 8 miles northeast of it. By 2100 the whole of the main road was firmly in German hands.

During the same night the regiments of the 255th Division occupied the other roads forming a triangle northeast of Maloryta. The enemy was trapped in the forest within this road triangle. The artillery of the 267th Division was then moved up and committed in support. On the evening of 25 June the enemy broke out toward Melniki, launched a surprise attack on two batteries of the 267th Division which had come up into the forest without proper security precautions, and wiped out most of the gun crews.

By 26 June the enemy's resistance seemed to weaken. Therefore, one battalion of the 455th Infantry Regiment was sent out from Lyakhovtse, on a broad front, to comb the forest toward the north. In the heart of the forest the battalion was suddenly attacked from all sides. The Russians had permitted the leading German elements to pass and then fired out of the trees. The battalion found itself in a critical situation and had to withdraw from the forest with very heavy losses.

On 27 June the entire forest was systematically raked with strong artillery fire; then the whole of the 475th Infantry Regiment was sent in to comb it from north to south. Enemy resistance was broken. Several thousand prisoners were taken and many guns and vehicles were captured. The Maloryta operation had required six days before it could be brought to a successful conclusion, and it was evident that a lesson had been learned.

At Pruzhany the mission of the 167th Division was to protect the highway to Ruzhana and Slonim against attack by Russian forces operating from the forest of Bialystok to the northwest and to guard the left flank of Fourth Army (map 3). Since the Russians were already very close to the highway, the entire 167th Division was ordered into the forest to relieve the situation quickly. Similar efforts at Ruzhana and Slonim soon caused a reduction in strength of the division at Pruzhany. In the initial action, the division veered off to the left of its route of advance in spearheads of regimental strength. There was no time for preliminary deployment.

As early as 26 June 1941, the enemy had appeared in the immediate vicinity of the highway near Pruzhany and had taken the road under machine-gun fire. Advance elements of the 167th Division were just barely able to halt the enemy west of Pruzhany. On 27 June the entire 167th Division was turned against him and was soon engaged in heavy defensive combat against a battle-worthy enemy division that emerged from the forest. In a stubborn battle to move the German flank protection farther to the west and away from the *Rollbahn*, the enemy was driven back to the edges of the forest. As soon as the first phase of this operation was successfully concluded, however, the lessons learned in the battle of Maloryta were applied and the pursuit of the enemy was carried no farther into the forest than was necessary to protect the *Rollbahn*. A planned regrouping of forces was then carried out. All elements of the division no longer required for purely defensive purposes, and a considerable portion of its artillery, were withdrawn while heavy machine-gun units were moved forward in their place.

In this case corps did not pursue the battle to a final decision, but halted the attack when its mission appeared to be accomplished. In the meantime weaker enemy forces were pressing toward the *Rollbahn* near Ruzhana, and signs of enemy activity were noticed near Slonim. Against these threats, elements of the 167th Division, mostly heavy machine-gun units, were at once moved by truck to Ruzhana and Slonim, where they succeeded in keeping the enemy off the *Rollbahn.*

The preceding example illustrates the German tendency to bypass forests wherever possible. As the commander of LIII Infantry Corps observed, "It is better to exercise discretion in the selection of objectives than to incur avoidable risks in forest fighting." This corps, at least, abandoned the practice of mopping up the enemy in forest areas and instead adopted as a standing operating procedure the practice of sealing off such areas. The following example shows how that technique had been elaborated and refined four months later as a result of combat experience.

Early in October 1941, LIII Infantry Corps was confronted with a security problem in the swampy region of Kletnya. Having been bypassed in the advance on Bryansk, the enemy in this area now constituted a threat to the left flank and rear of the corps, which was preparing to break through the fortified zone along the Sudost River (map, p. 18). Several weeks of position warfare allowed time for a thorough reconnaissance of the region, about which nothing was previously known. It was thickly forested, with fairly large, scattered patches of swamp, and was accessible along a number of narrow-gauge railways used for transporting peat. Numerous drainage ditches and high railroad embankments, some of the latter abandoned and with the tracks torn up, were characteristic of the region.

Part of the same swampy forest extended along the Bryansk-Roslavl *Rollbahn*, and here the rather weak front line security forces of LIII Infantry Corps were opposed by three Russian divisions. The possibility of a Russian advance from that area, supported by the enemy forces that were cut off in the swamps around Kletnya, represented a constant threat, especially as all combat troops were committed elsewhere.

On the basis of previous experience, LIII Infantry Corps decided that a mopping-up of the enemy in the Kletnya area would not be attempted. Instead, the whole region was to be sealed off. Two battalions of combat engineers and three of construction engineers were assigned this mission. Split into small task forces, each consisting of one construction battalion and one combat engineer company, these units proved capable of defending themselves as they performed their task. They blocked every stretch of dry ground around the edges of the swampy forest with mines and other obstacles and laid a triple mine belt along the *Rollbahn* opposite the enemy front. These mine fields were covered by engineer detachments placed at wide intervals. The mines, of which there were 40,000, proved very effective. Soon the construction troops were again free to take part in the preparations for the Sudost crossing while the combat engineer force, which remained in the swampy forest, proved of sufficient strength to keep the enemy in check. Two attacks in the direction of Kletnya by one of the three Russian divisions mentioned above bogged down in the mine fields, and the Russians were forced to withdraw. The enemy located in the swamps around Kletnya did not leave that area. As the corps' attack on Bryansk continued to make progress (see p. 19), the engineers could be withdrawn, and no further measures were taken there. The area was no longer of any importance.

Once the corps had succeeded in eliminating any active threat to its flank from the Russian forces located in the swampy forest, the terrain itself ceased to have any influence upon the over-all situation or the tactical operations of the corps. Neither the various arms—with the exception of engineers and construction troops—nor the air force or even the corps' supply operations were in any way affected. The engineers and construction troops alone had to cope with the difficult problem, both tactical and technical, which they proved capable of solving to complete satisfaction. It was a valuable experience to realize that the engineers could also be entrusted with minor combat and security missions. Schooled primarily for mobile warfare, they had to acquaint themselves now with the problems of defensive installations and the construction of obstacles. The chief engineer officer of LIII Corps personally conducted the tactical and technical operations in the swampy forest around Kletnya. He was also instrumental in obtaining the great number of mines needed.

The very first days of the campaign brought home the fact that the Red Army soldier bore little resemblance to the Russian soldier of 1914–17. The Bolshevist regime certainly understood how to imbue the Soviet soldier with a new spirit over the course of twenty years. This revelation was another of the surprises of the Russo-German war. The Russian soldier has always been brave and steadfast. But the new masters of Russia have succeeded in rousing the soldier from his passive stupor, in giving him a strong sense of responsibility toward state and nation, and even in turning him into a fanatic. The experiences and practices of the savage civil wars, in which the end justified any and all means, undoubtedly had much to do with molding the new Russian soldier and commander. In 1916 two German cyclists, one at the head of the column and one bringing up the rear, could easily march 500 newly captured prisoners 5 miles to the nearest enclosure. In 1941 this was out of the question. The Russian soldier of World War I was a tenacious fighting man but once the fortunes of war had turned against him, he resigned himself to his fate. The Russian soldier of 1941 kept right on fighting. By the fall of that year, the resurgence of dismembered enemy forces in forests and swamps behind the German front had assumed serious proportions. During the first part of the campaign, the German soldier was able to surmount such difficulties, but the longer the fighting lasted, and the poorer the quality of the ever-necessary replacements became, the greater were the effects of this situation.

The Soviet Union with its difficult terrain and its numerically well-nigh inexhaustible manpower potential, which moreover comprises an exceptionally high percentage of young age groups, can only be brought to its knees by a superlatively well-wrought instrument of war. That instrument Germany had at the beginning of the war, subject, of course, to the limitations that necessarily accompanied the rapid expansion of the army after 1935. But whatever the German units might have lacked in organic cohesiveness was partially set off by practical experience in four successful campaigns. Now that the sword had been forged and honed, the test lay in wielding it without blunting its fine edge. That its handlers failed to accomplish. Since the highest command steadfastly believed that the outcome of the campaign would be decided within eight or ten weeks, it was logical to hit the enemy from the very start with every ounce of Germany's military resources. But when the course of events had proved that belief a grave fallacy, the time had come for drawing the proper conclusions, not only with regard to the broader aspects of strategy but also with regard to economy of manpower. If defensive or offensive actions cost the Germans about the same toll of casualties as the Russians, the result in the long run had to be an exhaustion of Germany's war potential merely in terms of human lives. All the more inevitable was that final result if Germany's quantitative inferiority in manpower could not be offset by a qualitative superiority in matériel.

The Russians appeared to be well aware of these considerations. They chose for their most determined efforts swampy, forested terrain where superiority in matériel is least effective. For example, north of Gomel in the autumn of 1941 they attempted to establish a continuous line of strong points in the very thickest parts of the forest. Many of these strong points consisted of disabled tanks dug in and arranged for mutual support. They opened surprise fire on the approaching German infantry from thick underbrush or hollows in the terrain, and at first even assault units had great difficulty in getting close enough to deal with them. Finally, antitank weapons were brought up to close range and some of the defensive positions were captured. The fighting then resolved itself into numerous separate actions, and the enemy's resistance was broken in a comparatively short time.

Combat in forests has always been known as a costly type of warfare. The Russo-German war was no exception. The fact that the opponents clash in close terrain and come to grips in point-blank and hand-to-hand encounters leads by itself to numerous casualties. For this reason the very first day of the war saw the beginning of a search for new ways to support the infantry in the unavoidable forest fighting. Armor can nearly always be discounted as a supporting arm in dense forests; it is tied to the trails and open parts of woods. In addition, many of the large woodlands in Russia are interlaced with swamplands. As soon as armor encounters swampland, it is stymied and unable to support the infantry. Thus, only artillery and aviation are left to support infantry in forest fighting.

In the case of artillery, the difficulty lies in fire direction. In dense woods it is far from easy to identify and to fire at the enemy forward line. Even more difficult is the task of bringing effective fire to bear on the depth of the enemy position, because the targets are concealed from ground as well as air observation. An almost insoluble problem confronts the artillery if the attacking infantry has bogged down close to the enemy line; the enemy mounts a counterattack, and the battle see-saws back and forth in the woods. These are the very situations in which fire concentrations on the front line of the enemy would be particularly necessary and effective. In many instances, however, such a concentration is out of the question since friendly infantry stands in the zone of fire. Area fire was found to have the drawback of expending large quantities of ammunition without any certainty of success. As a result, many combined units resorted once more to observed and individually directed fire, although following a different procedure than in open terrain.

The difference in firing procedure lay in the fact that numerous forward observers accompanied the advance elements of the infantry and transmitted firing data to their batteries no longer by wire but only by radio. This procedure likewise has its disadvantages; it entails numerous casualties among forward observers and radio operators and causes the loss of a large number of radio sets. It furthermore requires extremely well-trained observers and gunnery personnel. Just as essential are immediately available reserves of observers and equipment. If these prerequisites are met, the infantry can be supported in a much more effective manner than in the case of area fire, which always remains more or less stereotyped in nature. The second procedure is much more flexible and more easily adaptable to the requirements of the moment. With proper training it is possible to coordinate the fire of an entire artillery regiment, and in a matter of minutes the fires of every battalion can be concentrated on one single point: a most effective procedure in preparing a breach of the enemy lines or in anticipating an imminent counterattack. If the artillery is properly schooled in its supporting mission, the infantry can even be withdrawn a short distance before fire for effect commences on the enemy forward lines. So long as infantry and artillery function as a smoothly coordinated team, the enemy will be unable to utilize that opportunity for a follow-up thrust or an advance of his forward line.

Successful artillery support in forests requires intensive special training and presupposes certain favorable conditions. The terrain must allow a reasonable minimum of mobility to the guns, and survey personnel must have time to plot the positions of the guns precisely before they are required to fire. But since German troops avoided combat in forests whenever possible, the typical forest operation was a mopping-up of dispersed Russian forces. This was normally carried out as speedily as possible by units in the strength of regimental combat teams. If the forests were at all swampy, the artillery had great difficulty in keeping up with such units. Moreover, in this type of fighting, artillery support was needed quickly or not at all. It was seldom possible to locate the target and begin effective fire in time. From repeated experience the Germans came to the conclusion that, except in unusual circumstances, artillery in thick forests was just an impediment and should be left behind. They much preferred heavy mortars such as the Russians used in large quantities and with great effect. These could be hand-carried over even the most difficult terrain and, being comparatively short-range weapons, they involved only minor problems in fire control.

The infantry's task in forest fighting was considerably facilitated whenever the artillery was augmented by combat aviation support. As spotty as reconnaissance data were and had to be by the very nature of wooded terrain, particularly in the case of an enemy as proficient in camouflage as the Russian, the commitment of combat aviation against the depth of enemy positions produced excellent results. To be sure, supporting aircraft must be able to identify with absolute certainty at least the friendly if not the hostile forward line of a forest position. Mere designation of that line on the map is not enough; the bombardier must have a visual reference point. Thus, the only sure-fire methods of identification are ground signals, magnesium flares, incendiary markers, smoke pots, signal beacons, and similar visual aids.

As long as the Stuka units had all their excellently trained personnel, they could be committed for knocking out enemy resistance in the forward lines of forest positions. However, the use of Stukas for that purpose necessitated minute briefing and target designation, as well as the arrangement of a precisely timed plan of attack. A serious disadvantage lay in the fact that the ground forces could not establish radio contact with the approaching Stukas. Nevertheless, the Stukas performed excellent service. Friendly infantry usually approached to within 100 yards of the enemy forest position. The Stukas then laid their heavy-caliber bombs with astounding precision directly in front of the infantry. The assault got under way as soon as the last bomb had hit the ground.

Relatively heavy-caliber missiles were employed against forest positions. In the spring of 1942, for example, Stukas dropped 1,000-pound bombs on the Soviet main line of resistance in the woods south of Lake Ilmen. The whine of diving Stukas, the explosions of heavy bombs, and the crash of toppling trees had a great moral effect, particularly if hostile forces were experiencing their first Stuka attack. The physical effect, on the other hand, was negligible in most instances. An enemy entrenched in a forest position usually suffered few casualties. Even the moral effect decreased noticeably as time went on. Thus, in the final analysis it was still the infantryman who bore the brunt of the forest fighting.

The same was true in the case of combat in swamps. The effect of artillery fire is largely nullified by swampy soil. Aerial bombs must be equipped with extension-rod fuzes (*Vorsatzzuender*) to produce any sort of result. Above all, however, combat in swamps puts the infantry to unusually severe physical tests. In summertime the men constantly live among dampness and moisture. Boots and uniforms begin to rot; myriads of mosquitoes never cease to make life miserable; drinking water is a rare and precious commodity; proper body hygiene is impossible; epidemics of diarrhea, dysentery, and typhoid spread like wildfire. Among the principal reasons for the high casualty rates in swamp fighting is the impossibility of digging field fortifications. Every cut of the spade immediately fills up with water. Cover can only be built with logs, which frequently have to be hauled a long distance, and with sandbags.

In wintertime some of the swamps freeze over and others remain open. The swamps around Lake Ilmen normally remain open during the winter, or so the inhabitants claimed. Toward the end of December 1941, however, not only the swamps but also the lake itself was covered with a thick layer of ice. The Russians attacked across Lake Ilmen and the swamps and split the German Sixteenth Army into two

MACHINE GUN SQUAD goes into position in swampy forest.

parts, the larger of which eventually became completely encircled (the Demyansk pocket).

The influence of forest and swamps on military operations in Russia plays a crucial role particularly in one respect; namely, partisan warfare. Partisan warfare dates back to the earliest days in Russian history and appears to be a universal characteristic of all Slavic peoples. The Serbs and Bulgars take to the hills, as during the years of fighting by the Comitadjis in Macedonia; the Poles and Russians take to the forests. An account of partisan organization and tactics in the USSR lies outside the scope of the present manuscript. Those subjects are properly part of a study on the Soviet conduct of operations. *See* CMH Pub 104–12, *Russian Combat Methods.*

Only a future history based on irrefutable primary sources will be able to assess the long-range effect which the painstakingly organized partisan operations had on the German campaign against Russia. The immediate repercussions were serious. Partisan warfare was a festering, ever-spreading cancer. Anyone flying over occupied Russia in 1943 or 1944 got an excellect picture of how far the disease had spread. Pilots were issued maps on which partisan-infested areas were circled in red. There was an amazing number of these red circles. Within each circle was a number, stating the minimum altitude for crossing the particular area. Rovno, Borisov, Gomel, Bryansk, Vinitsa, and other places were marked 2,500 meters (8,300 feet), which meant that partisans in those territories had antiaircraft artillery.

In the wooded and swampy regions between Velikiye Luki and Lake Peypus the guerrillas conducted full-fledged conscription drives behind the German front. Under cover of darkness the conscriptees were funneled through gaps in the German front, which north of Kholm consisted only of strong points, and subsequently received military training in the Soviet rear areas. In other regions the partisans are said to have conducted winter maneuvers, including firing practice with live ammunition by heavy weapons and even artillery.

Antipartisan operations required a substantial number of forces that otherwise might have fought at the front. Guarding rear lines of communication, particularly railroad lines and operating facilities, airdromes, ammunition and ration dumps, etc., absorbed hundreds of thousands of men. Despite the guard details, however, the rail lines were blown up with clocklike regularity. At times the demolitions assumed such proportions that the supply of the front became seriously endangered. There were times when night traffic on vital feeder lines had to be completely shut down. The trains, with sand cars ahead of the locomotives, ran only during daytime, one train following within sight of the preceding one. Of course, the trains did not run at all if, despite increased security patrols, the rails had been blown to pieces during the previous night. In several instances of impending German offensives, a host of partisan units launched joint, large-scale operations against rights-of-way and other railroad facilities. Prior to the German Kursk offensive in the summer of 1943, for example, the strategic Desna bridge at Bryansk suddenly blew sky-high, and a series of unusually numerous demolitions, set off as if by prearranged signal disrupted the Roslavl-Bryansk railroad.

A field day for the partisans came when the Germans had to withdraw. The German units had to fight on two fronts. The partisans anticipated the routes of withdrawal and systematically destroyed every bridge in the rear area. This wholesale destruction of bridges gave rise to extremely critical situations. With the main enemy hard at their heels, German forces were frequently compelled to send flying columns to the rear for the purpose of capturing the destroyed bridge sites in enemy hands and building new bridges.

Among the multitude of difficulties which partisan warfare caused in matters of personnel and matériel, one salient outgrowth of that particular struggle must not be overlooked. Today, there is no longer any need to prove that the partisans fought their war with unusual cruelty. And if warfare in the East became more and more bitter, the partisans contributed no small share toward that end.

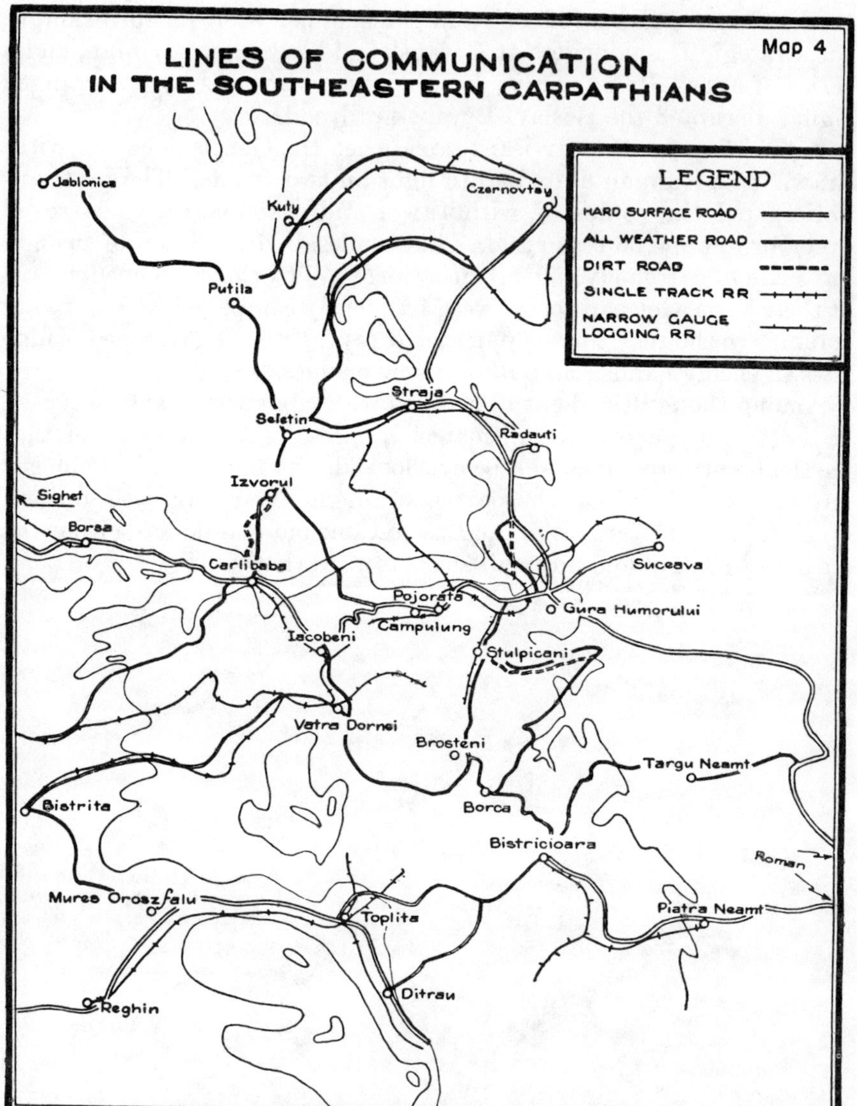
Map 4
LINES OF COMMUNICATION
IN THE SOUTHEASTERN CARPATHIANS
LEGEND
HARD SURFACE ROAD
ALL WEATHER ROAD
DIRT ROAD
SINGLE TRACK RR
NARROW GAUGE
LOGGING RR
Jablonica
Kuty
Czernovtsy
Putila
Straja
Seletin
Radauti
Izvorul
Sighet
Borsa
Carlibaba
Suceava
Pojorata
Campulung
O Gura Humorului
Iacobeni
Stulpicani
Vatra Dornei
Brosteni
Targu Neamt
Bistrita
Borca
Bistricioara
Roman
Mures Oroszfalu
Toplita
Piatra Neamt
Ditrau
Reghin

SECTION VI

DEFENSE AND WITHDRAWAL IN HEAVILY WOODED SUBALPINE TERRAIN

(Fighting in the East Carpathians in 1944)

Military Topography of the East Carpathians

Although the East Carpathian Mountains are outside of Russia proper, they were the scene of Russian operations in both world wars and part of the territory has now been annexed by the USSR. The geographic location and terrain features of these mountain ranges give them a certain strategic significance, especially today when the whole area constitutes an outer defense zone for the Soviet Union. Moreover, the Carpathians have local political significance as a region where several national frontiers meet.

Wherever there are heavily forested mountains of medium height, the defender is afforded a variety of tactics in conducting his operations. Active defense, delaying actions, the establishment of a system of protected road blocks, and flanking attacks against enemy forces bypassing the mountains or against front lines anchored upon them are some of the methods he may employ. On the other hand, the attacker may conceal his forces during the assembly prior to offensive operations. These conditions apply not only to the Carpathians but also to similar terrain in other parts of the world.

This section is concerned only with that area in the southeastern Carpathians north of Piatra Neamt-Kuty, west of Chernovtsy, and east of Sighet-Bistrita-Mures Oroszfalu (map 4). This region is characterized by a scantier population and fewer communications facilities than similar areas in Western Europe. Here the Carpathians show the characteristics of a subalpine mountain range: They are covered with tall stands of timber extending clear to the mountain tops. The average elevation of the mountains is 2,000 to 3,500 feet in the east, 4,000 to 5,500 in the central sector, and 3,500 to 7,500 feet in the west. The Caliman Mountains southwest of Vatra Dornei, the Rodnei Mountains south of Borsa, and the Czarnohora east of Sighet are the only massifs which rise to approximately 7,000 feet and have the alpine features of rocky, bare peaks.

The valley of the Bistrita extends from Carlibaba (southwest of Vatra Dornei) via Vatra Dornei toward Piatra Neamt. Along this general line, the region is divided into two parallel mountain ranges stretching from northwest to southeast. A defender facing east will establish his main line of resistance along the course of these mountain ranges. The depth of the defense sector is therefore considerable, a factor which accounts for its particular strength. An attacker who intends to penetrate the eastern Carpathians from the east will not direct his strategic main effort toward this area in which he has to overcome two successive mountain ranges. He will atempt a thrust farther to the north, for instance, across the well-known Tartar Pass near Jablonica toward the area northeast of Sighet, where the mountain range is shallower from east to west.

The network of highways, dirt roads, and railroads is adequate; but a permanent road maintenance service, as well as snow clearing detachments in winter, is needed for all roads, especially where they extend through defiles and across passes. The main traffic arteries and through roads can be kept open even in winter. Most of the bridges are constructed of wood. Half-tracked and armored vehicles must bypass them and use the fords which have firm, stony bottoms and can be found in the vicinity of every bridge. The railroads are not capable of very high performance; their daily average capacity varies between 8 and 12 trains. The numerous narrow-gage logging railroads may be used for supply and minor troop movements.

The population is composed of very frugal farmers, lumbermen, and herdsmen in the rural districts and tradespeople in the cities. As a frontier people who have been forced to change their political allegiance several times during the last 30 years, they distrust all strangers.

The climate in this elevated forest region is healthy throughout the year. Warm clothing should be kept on hand at all times, even during the summer, because the nights are generally cool, especially at high elevations. The winters are usually very cold and snowy; but if the troops are properly equipped, climatic conditions will be tolerable.

The Tactical Situation in Spring 1944

In March 1944, the Russians made an offensive thrust toward Bessarabia and opened a gap between Army Group South and the German army group adjacent to the north. The Soviets now faced the extensive forests of the Carpathians which, in parts, were still covered with snow.

A temporary defense of the Carpathians was attempted by commiting replacement units hastily brought up from the rear, Romanian formations, and so-called Carpathian battalions. The latter were improvised units composed of miscellaneous troop elements which had become detached in the course of the withdrawal. All these troops were inexperienced in forest and mountain fighting, and lacked the special equipment needed for this type of combat. If they succeeded in stopping the enemy advance into the mountains, it was only because the Russians initially employed weak forces and the wooded terrain was particularly suitable for defense. For the moment the Russian command was intent on capturing Bessarabia and showed no intention of penetrating the eastern Carpathians with strong forces. In any event, these mountains had assumed a singular importance and significance in the eyes of the German command. The Carpathian front, even though it was only weakly held, constituted a potential threat to the flank of the Russians whose main forces had penetrated northern Bessarabia. The creation of that front forced the enemy to provide flank protection.

At the end of April 1944 the general in command of XVII Infantry Corps was assigned to that sector of the Carpathian front which extends from north of Piatra Neamt to south of Kuty. The corps was subordinate to Eighth Army. A few weeks later the replacement units and so-called Carpathian battalions were relieved by the 3d Mountain and 8th Jaeger [Light Infantry] Divisions. These two units, which had hitherto performed well in combat against the Russians and had received appropriate training and equipment for forest and mountain fighting, subsequently proved very effective.

The Defensive Mission of XVII Infantry Corps

Upon taking over the sector, XVII Infantry Corps was given the following mission:

a. To secure and defend the gateways to the Carpathians against enemy reconnaissance and attacks with limited objective.

b. To defend the passes situated behind the gateways should the enemy launch an offensive with the objective of penetrating the Carpathians. There are six well-defined passes in this area. They are located northeast of Borca and Borsteni, southeast and east of Vatra Dornei, northeast of Iacobeni, and east of Izvorul (map 5, p. 48).

To accomplish this mission, the corps made the following dispositions:

a. Security and defense of the gateways: The main defensive efforts were to be made at the approaches to the Carpathians. Openings for flanking movements were to be eliminated. Control of the intermediate terrain was to be achieved by establishing a system of strong points from which intensive reconnaissance patrols were to be initiated. Special security detachments were to protect supply routes. Finally, strong reserves were to be held back at central points to eliminate delays in shifting them to danger spots. Each division was to constitute a reserve of two to three battalions and provide it with all the truck transportation available.

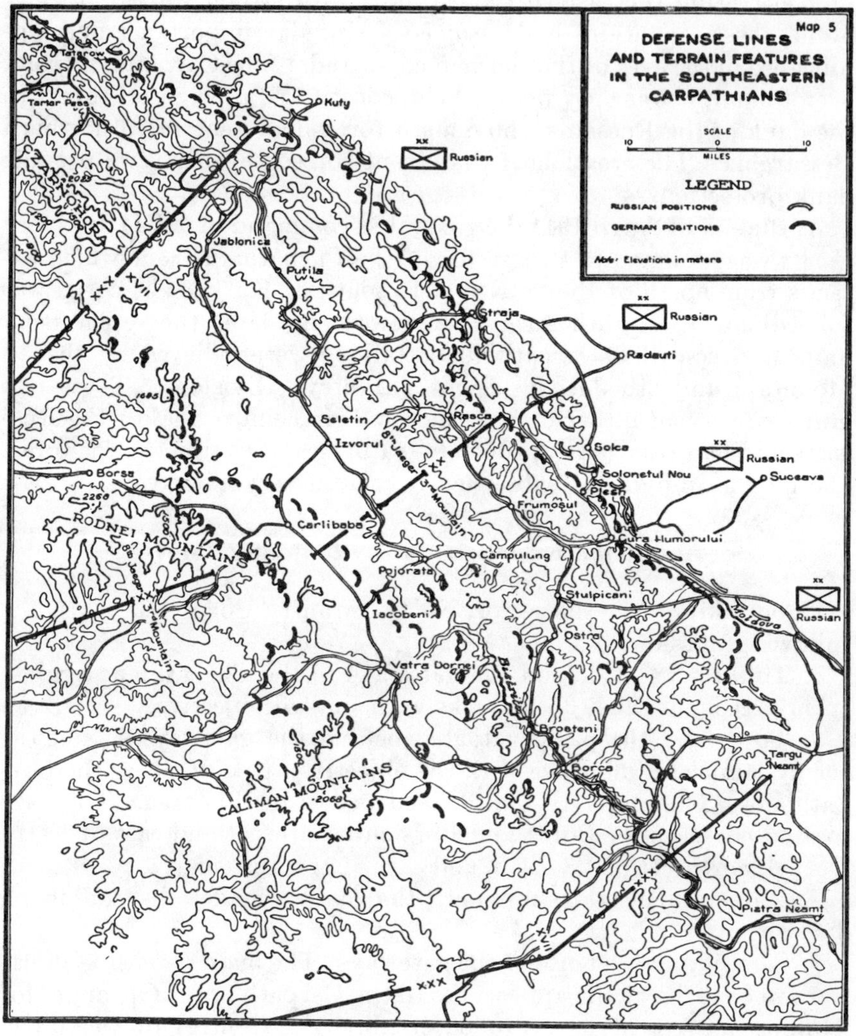
Map 5
DEFENSE LINES AND TERRAIN FEATURES IN THE SOUTHEASTERN CARPATHIANS
SCALE
10 0 10
MILES
LEGEND
RUSSIAN POSITIONS
GERMAN POSITIONS
Note: Elevations in meters
Tatarow
Tartar Pass
Kuty
Russian
ZAROHORA
Jablonica
Putila
XXX XVII
Straja
Russian
Radauti
1685
Seletin
Rasca
Izvorul
8th Jaeger
3rd Mountain
Solca
Russian
Suceava
Solonetul Nou
Borsa
2268
Frumosul
RODNEI MOUNTAINS
Carlibaba
Gura Humorului
Campulung
Pojorata
Stulpicani
Russian
Iacobeni
Moldova
Ostra
Vatra Dornei
Bistrita
Broşteni
Borca
Targu Neamt
CALIMAN MOUNTAINS
2068
XXX
Piatra Neamt
XXX

b. Defense of the passes: The defensive forces were to be concentrated at the passes. Strong reserves were to be held back for counterattacks. Active reconnaissance patrols were to observe the intermediate terrain.

This mission was to be accomplished by two German divisions and Romanian units to a total strength of one and a half regiments. These were all the forces available along a front extending over 80 miles as the crow flies. At first, the Russian troops facing these forces amounted to approximately three and a half divisions which were also deployed for defense. Ordinarily, much stronger forces would have been needed for such a wide sector; for the time being, however, no Russian offensive was anticipated along this front. Moreover, the terrain favored the defender. Thus, from the point of view of Army Group South and Eighth Army, the strategic significance of the Carpathian front was twofold: It saved forces for the main defensive effort in Bessarabia, where the next major Russian offensive was expected, and it contained enemy forces.

The alinement of the positions destined to serve the defense of the passes was relatively easy to determine because most of the Carpathian passes are situated in very dominant positions. Fortifications erected during World War I were discovered along the passes 3 miles northeast of Iacobeni, twelve miles southeast of Vatra Dornei (on the road to Brosteni), and in the intermediate mountains. These were integrated into the course of the main line of resistance and incorporated into the construction program.

Openings for outflanking the positions were eliminated wherever possible. This problem deserved special attention in the area around Borca and Brosteni, where the extremely important communication road in the Bistrita Valley was exposed to being cut off by an enemy thrust from Stulpicani via Ostra and the mountains. Another danger area was east of Vatra Dornei, where a surprise thrust from the east might eliminate this crucial communications junction. Finally, the area around Izvorul and north of Rasca was particularly vulnerable to premature enemy thrusts toward the Pojorata-Seletin-Straja road or in direction of Carlibaba toward the Borsa-Iacobeni highway. Subsequent combat actions during the retreat corroborated this estimate.

Defensive positions were anchored on natural obstacles which were difficult to penetrate or surmount. This resulted in savings of manpower, troops, and time required for the construction of fortifications. The forward area of the western sector in former Hungarian territory had all the necessary prerequisites. Here the positions were supported by the Caliman and Rodnei Mountains and by the southern spurs of the Czarnohora.

Suitable lines of supply and communication leading to the positions were selected and provided with all possible protection. This was important because no position in a forest or on a mountain can be held without such lines. Even tactical considerations had to be subordinated to this factor.

Field-type fortifications were erected in the eastern and central part of the sector; in the western, most of the construction work consisted of improving the permanent fortifications along the passes and the positions barring the valleys, with field-type structures filling the gaps Priorities in the construction program were determined in accordance with the prevailing tactical situation and the terrain conditions. The defensive positions along the passes and across the valleys were ready within a relatively short time. The main emphasis of the construction effort was subsequently placed on strengthening the lateral positions on the adjacent heights. The highest priority was given to clearing fields of fire.

The Defense of the Carpathians

During the entire phase of position warfare, which lasted from April to mid-August, and the subsequent withdrawal up to the beginning of September 1944, the terrain had the anticipated effect on the conduct of operations. The eastern security line in the area around the gateways was considered capable of defense, thanks to the relentless construction efforts of the combat troops. In the intermediate terrain, the sparsely occupied line of strong points, with intervals up to 2 miles separating the individual installations, continued to be of value only for purposes of surveillance and security. Reconnaissance, security, and defense were conducted very actively during the period of position warfare.

Friendly and hostile reconnaissance, notably long-range patrols across the lines into the rear areas, were conducted with particular intensity in the northern half of the corps sector. There, the terrain was very close and the strong points were separated by wide intervals. German reconnaissance and combat patrols penetrated as far as the eastern foothills of the Carpathians, brought back useful intelligence, and disrupted enemy supply and communication very effectively. The Russian patrols were unable to penetrate equal distances into the German rear areas. In this terrain, the specialized troops of the 3d Mountain and 8th Jaeger Divisions proved superior to the soldiers of the Russian rifle divisions. Romanian soldiers were used as guides by the Germans because they were familiar with the terrain.

Pack columns with mules were used to bring up supplies to the troops in the strong points on top of the mountains. Since only the 3d Mountain Division had mules, some of its animals were turned over to the 8th Jaeger Division. The supply columns took security measures on their way to the front and, when passing through particularly dangerous areas, were accompanied by security detachments provided by the combat units.

In mid-August 1944 the Russians began their large-scale offensive in Bessarabia. Because of the lack of strategic reserves and the surprise caused by the defection of the Romanian units, that offensive finally led to the conquest of Romania by the Russians. The major part of the Sixth German Army, which was committed in the eastern sector of Bessarabia, was cut off and encircled. The Eighth Army, in spite of the defection of the Fourth Romanian Army, succeeded in pivoting its right wing from the western part of Bessarabia toward the Carpathians. The army was able to reestablish a continuous front in conjunction with XVII Infantry Corps which was holding the Carpathian front on its left. It effected a very difficult but nevertheless orderly retreat across the Carpathians and held off the Russians, who were following closely although they were hampered in their conduct of operations. In this instance, the Carpathians had, therefore, assumed once again a great strategic significance by virtue of their geographic location and terrain features.

When the Russian offensive began, the Romanian units under XVII Infantry Corps deserted as they did elsewhere. The defection of the Romanian regiment deployed along the Moldava southeast of Gura Humorului opened a dangerous gap in the vicinity of this extremely important gateway into the mountains. However, reserves of the 3d Mountain Division were moved up quickly and formed a defense line southeast of Stulpicani and south of Gura Humorului, and they prevented the Russians from advancing in the direction of the Campulung-Gura Humorului road. Eighth Army issued an order widening the XVII Infantry Corps sector for the duration of the withdrawal. It now reached up to the Targu Neamt-Bistricioara-Toplita-Mures Oroszfalu road; two divisions held a front of 100 air miles (map 4).

The Withdrawal

Two factors determined the plan the corps drew up for the retreat: the available roads and the defensive positions protecting them against enemy interference. Though the latter had been fully reconnoitered, they were only partly constructed.

Russian intentions could be recognized at an early stage. One division was committed at each gateway as a first assault wave; thus, the corps was faced by four to five divisions. The main attack aimed, as expected, in the direction of the Bistrita Valley, at a point between Bistricioara and Vatra Dornei. This constituted the most obvious plan of operations considering the terrain conditions. In the southern and central sectors, the direction of the highways and alternate approach roads was favorable; in the northern sector, most roads, except for the main routes of advance, cut right across the direction of attack. In addition, the main routes of advance in the southern and central sectors were closer together which guaranteed better opportunities for mutual cooperation. A thrust following the course of the Bistrita would have presented the enemy with too many difficulties because of the defiles and defensive positions northeast of Vatra Dornei.

Both German divisions exploited the favorable terrain of the mountain forests and employed active defense tactics in all instances. Although the retreat came as a complete surprise and was fraught with numerous difficulties, its first stage was carried out without disrupting the continuity of the front. In view of the small forces available, this was a notable performance. Its successful accomplishment was the prerequisite for the continuation of the retreat.

The corps held the positions west of the Bistrita Valley and northeast of the Vatra Dornei-Carlibaba road for several days. It then withdrew to the well-prepared Hungarian border positions without giving the enemy any opportunities to interfere with its movement. Tactical and strategic considerations motivated the accelerated withdrawal of the units adjacent to the south. As a first result, the right wing of corps was withdrawn to the exit from the Carpathians east of Mures Oroszfalu. The 3d Mountain Division once again had to engage in very heavy fighting in this sector. During the further course of the retreat of the German Eighth Army, XVII Infantry Corps withdrew entirely from the Carpathians in mid-September 1944.

Summary

The decisive superiority of specially trained and equipped mountain divisions over conventional rifle divisions, when both are employed in mountainous terrain, has been amply demonstrated in the preceding pages. The tactics of defensive mountain warfare as illustrated here are fairly well known. An aggressive attitude on the part of both command and troops is of primary importance. Otherwise the difference from conventional infantry tactics is a matter of emphasis rather than of basic principles. Patrols, for example, are a feature of all infantry tactics, but close mountainous terrain makes long-range reconnaissance patrols deep into enemy rear areas a practical possibility. Aggressive and well-trained troops can exploit this opportunity with good effect, especially if they have been trained in the advance recognition of enemy intentions.

Prepared defenses along mountain ranges and the use of specialized troops can effect important economies in manpower; defensive sectors may even be twice as wide as those assigned to units in normal terrain. In the Carpathians, German units were under strength and were not given sufficient time and matériel to prepare satisfactory defensive positions. Nevertheless, they were able to hold out until events in other areas dictated a strategic decision to withdraw from the mountains.

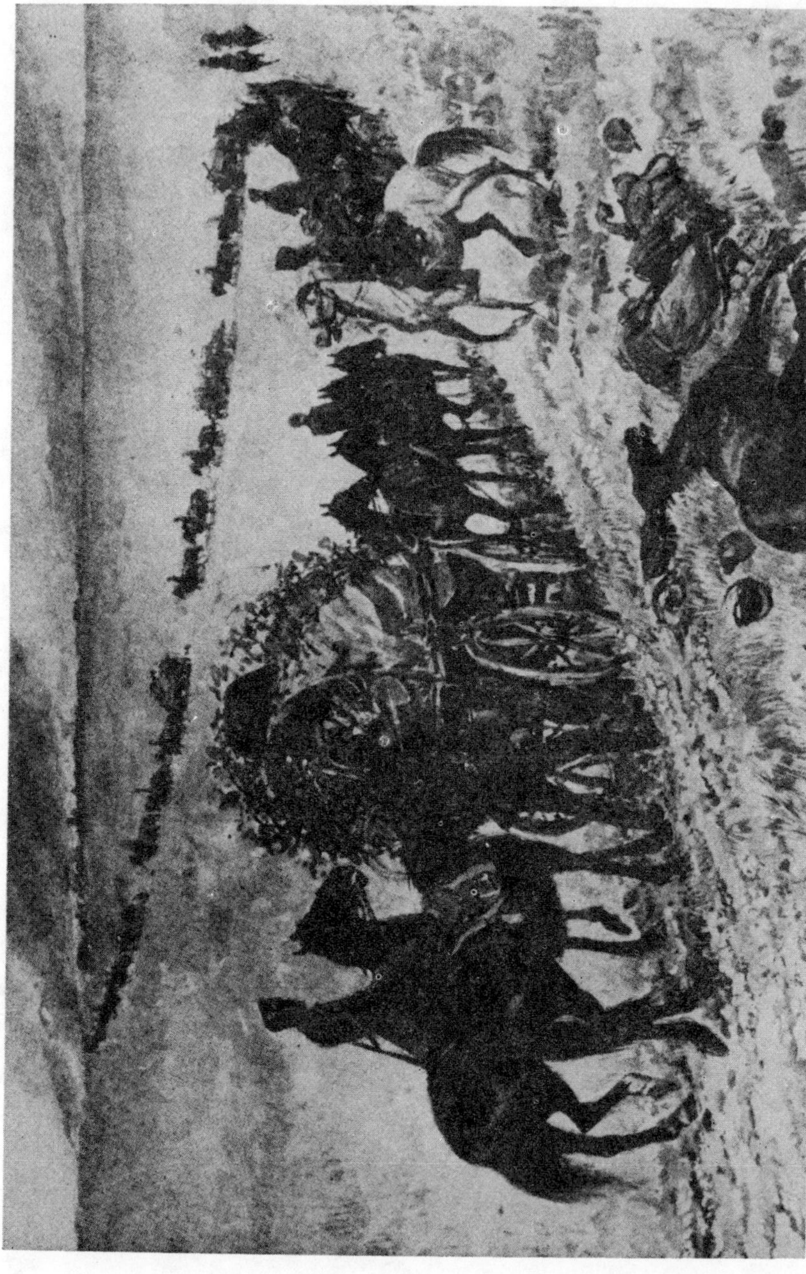

GERMAN SUPPLY COLUMN in the Nogayskaya Steppe.

SECTION VII

COMBAT IN THE STEPPES

The term steppes implies untilled wasteland and can be applied to large areas of the USSR, especially in Asia. In the Middle Ages the whole of southern European Russia, including the Ukraine, was steppeland. Since that time the Ukraine has become the granary of Russia, and all the steppelands west of the Volga are being used more and more for agricultural purposes. The Soviet Union has accomplished much in this field of reclamation.

During World War II, the German forces penetrated only the western edges of this area. There were extensive military operations on the Nogayskaya Steppe, between the lower Dnepr and the Sea of Azov, but this is now a flourishing agricultural region that hardly merits the name of steppe. The steppes on the Crimean Peninsula were found to be interspersed with cotton plantations and even fruit orchards, and the steppes along the Don have also been partially reclaimed. Of all the territory reached by the Germans during World War II, only the Kalmyk Steppe, between the Don and the Terek Rivers, has retained the character of treeless grazing land.

During the temperate seasons of late spring, summer, and early fall, the steppe is the ideal battleground for armored and motorized units. The vast flatland is accessible in all its parts and harbors only one natural obstacle—the so-called Balkas. These are ravines, in many instances with high and steep slopes, overgrown with brush and thickets. But the Balkas can be reconnoitered and avoided. The much more serious threat in the steppe comes from hostile aviation because the Balkas, some isolated clumps of trees, and a few human settlements are literally the only features to afford cover. During the early fighting in the Russian steppes, the Germans had air superiority. At that time the lack of cover placed the withdrawing enemy forces at a decided disadvantage.

Later the tables were turned, and the disadvantage was felt just as keenly on the German side. In its retirement through the steppes of the Kerch Peninsula in the spring of 1944, V Infantry Corps had not one single tank. The steppeland on that eastern landspit of the Crimea is hilly, with full-fledged mountains rising in the immediate vicinity of the Black Sea and the Sea of Azov. The weak elements of German combat aviation were committed against the Russian breakthrough on the narrow isthmus that joins the northern Crimea with the mainland. As a result, the very strong Red Air Force had absolutely free rein over the peninsula. The Soviet air and armored forces arrived at a very simple division of labor: while the tanks proceeded to put German infantry out of action, the air force swooped down on the prime movers of heavy weapons and artillery, as well as on vehicles in general. The outcome was that V Infantry Corps had soon lost the bulk of its artillery. In that completely open terrain, every horse- or tractor-drawn gun was plainly visible over a considerable distance. Horses and motorized equipment were blown to bits by strafing aircraft, and the last few prime movers only succeeded under cover of darkness in towing away some of the guns. The rest had to be blown up. A few squadrons of German fighter aircraft could have saved a substantial part of the artillery, prime movers, and motor vehicles from destruction at the hands of Russian combat aviation.

The school solution, of course, would be to conduct major retrograde movements across steppeland only at night. In theory that solution sounds fine but in practice it is only possible if the defender succeeds in holding his position throughout the entire day until nightfall. If the attacker gains ground during the day, or if a breakthrough becomes imminent or actually succeeds, the defender is unable to avoid large-scale movements during daylight hours. If the retiring defender lacks equality of airpower, not even the best-organized dispersal of his units or the most artful camouflage will save him from his fate. The individual can always conceal himself with steppe grass, the sole but abundant camouflage material in that terrain. A moving gun, a prime mover, or a column of trucks carrying reserves, however, cannot be concealed in the open steppeland, not even with a multitude of camouflage nets.

One of the major problems during combat in the arid steppe is a proper supply of drinking water. Each unit must have its own water truck. No less important are a number of other factors peculiar to that type of terrain. Except for the Balkas, the steppe offers no natural cover against air attack or artillery fire, so that bombs and shells have a particularly serious fragmentation effect. During prolonged halts motor vehicles had to be dug in. But some of the most severe hardships in the steppe result from the shortage of wood. Without wood, no fires can be built in the field kitchens and no shelters constructed. In the steppe, wood becomes a highly treasured commodity, above all in the winter.

Winters in the steppe are rigorous. Neither forests nor mountains break the icy east wind or bank the snow. The wind whips the snow across the plains and into the only large depressions, the Balkas.

The Balkas, however, are the only places that offer shelter for men and vehicles. The result is a never-ending struggle against the drifting snow. Without well-regulated and ample shipments of heating fuel, a military force cannot survive a winter in the steppe. White camouflage clothing and white paint for vehicles are indispensable.

The thaw radically changes the face of the steppe. The Balkas churn with swelling streams. Small depressions are transformed into ponds, and large areas of flatland are covered with water. The thawed ground becomes soggy; the muddy season begins. For a period of weeks neither troop convoys nor individual vehicles can move for any but the shortest distances. Only tracked vehicles are able to drag themselves through the mire.

Summertime in the steppe exposes a military force to yet another hazard—deliberately set fires. In the summer of 1942 the Russians repeatedly put the dry steppes to the torch. Fortunately, the fires never assumed such proportions that the troops were in bodily danger; they only were forced to a rapid change of position. Nevertheless, the flames destroyed ammunition and equipment that could not be promptly evacuated. There is room for doubt, though, whether the Russians ever were bent on starting full-fledged conflagrations, even at places at which their aircraft dropped phosphorus. These scatterings of incendiaries were more in the nature of local, uncoordinated nuisance raids, but the possibility of setting large-scale steppe fires with wholly different results is not to be ignored.

Finally, the steppe is ideally suited for parachute and air landing operations on even the largest scale. The terrain offers completely unobstructed landing facilities for glider and transport aircraft. But neither the Germans nor the Russians took advantage of that opportunity during World War II.

SECTION VIII
SUMMARY AND CONCLUSIONS

Never in history has a one-sided attack from the West succeeded in subjugating Russia. The record cites the abortive attempts of the Poles, the Swedes, Napoleon, and lastly the Germans. Are these failures to be attributed to the terrain of western Russia, the tough fiber of the USSR's inexhaustible manpower potential, or perhaps the faulty strategy of the invader from the West?

The Teutonic Knights, the Poles, and the Swedes waged most of their wars against Russia during winter, the season in which ice and snow neutralize many of the natural obstacles. Every one of the campaigns follow the monotonous pattern: The attacking armies achieve initial successes (the Poles carried Moscow), the Russians withdraw into the interior of their domain, and finally they crush the attacker by sheer weight of numbers.

Napoleon postponed his campaign against Russia until summer; he overlooked the fact that the Russian summer is short. Like his predecessors, he scored brilliant successes at the outset; but before he reached Moscow his army suffered extremely high casualties from disease and exhaustion. He lost half of his horses, not as the result of enemy action, but because of the heavy footing in the rain-soaked countryside. It is no secret that Napoleon fervently wished for the decisive battle. The Russians refused to do him that favor; they retired into the depth of their country. Those elements of the Grand Army that finally did reach Moscow were no longer a match for the ever-increasing numbers of the enemy. The deciding factor, however, was not the huge losses of men and matériel in the Russian terrain, but the fact that Napoleon had not succeeded in annihilating the Russian Army. This army and the early winter delivered the death blow. Napoleon failed because all he had left in Russia by October 1812 was a shrunken army in a burned-out capital and an overextended and unsafe supply line. Wherever he looked he saw nothing but enemy forces.

These dangers of the vast Russian land masses were to be countered in 1941 by the German assault along a wide, continuous front from the Black Sea to the Barents Sea. German strategy promised to eliminate the factors responsible for the downfall of Napoleon: the threats to the flanks and rear inherent in the immense territory of Russia. And yet, the invasion ended in complete failure.

The Russian terrain had profound effects on the German Army. Although casualties from sickness and exhaustion were astonishingly few, the losses in matériel were serious. Heat, dust, poor roads, marshy terrain, and too few stops for repairs wreaked such havoc with German armored and motorized equipment that no amount of subsequent effort could undo the damage. The theory that all limitations of distance have been conquered by the internal-combustion engine proved to be false. Difficult terrain imposes a performance limit on tanks and motor vehicles in general; thereafter they must be overhauled or replaced. Since German industry was not equal to that task, the German Army, whose equipment was much too heavy for warfare in Russian terrain, lost more and more of an essential part of its striking power; namely, its mobility. The terrain took its toll not only of the motorized equipment but also of the horses. As early as 1943 a distinction in status had to be made between mobile divisions and those that were virtually without transportation.

Despite the ravages of the terrain, however, the German Army—in contrast to the Grand Army of Napoleon—reached the Don and the outskirts of Moscow and Leningrad with its striking power unshattered and its morale unbroken. It therefore appears that the influence of terrain upon the conduct of operations was *not* the decisive factor that brought about the defeat of the German forces in Russia. After this observation, it hardly seems profitable to pursue this subject any further, all the more since the controversy about the primary cause of Germany's defeat in the struggle with the Soviet Union is still far too intimately affected by recent events to allow a lucid and wholly objective judgment. Such judgments must be derived from exhaustive study of all the pertinent data in the light of historical perspective.

One preliminary phase of such a study would be a critical examination of the grand strategy underlying the Russian campaign, in itself a tremendous undertaking that cannot even be outlined here. But a few of the themes to be developed in the course of such an investigation can at least be mentioned, and these inevitably lead back to a consideration of terrain, but on the level of strategy rather than operations.

Many another country would have collapsed under the blows that the German Army dealt Soviet Russia in 1941. Many serious errors were committed by the German command, the most serious of them perhaps the supposition that the entire structure of the Soviet regime would topple after the first defeats of the Red Army. Like Napoleon before them, the Germans were unable to destroy the Russian Army. The Russian Colossus could afford to withdraw his troops and sacrifice whole regions. In the final analysis, the loss of the Caucasus and the Don severed only a part, even though a valuable part, of the western ramparts of the Russian heartland.

The very moment, however, in which it became obvious that the attempt to crush the Russian armed forces had failed and that the Red Army was about to continue the war from the depth of its motherland, the Russo-German war had entered a new phase. The view was widely accepted, and not based on hindsight either, that the time had come for a radically new strategy: to hold as many of the occupied territories as necessary, to take full advantage of seasonal difficulties, to allow the Russians to butt their heads against deeply echeloned defensive systems, and, finally, to deliver the counterblow. Whether such a conduct of operations was feasible at that particular time is difficult to determine today; Germany was at war with almost the entire world, and was fighting not only in Soviet Russia but also in North Africa.

Was it the Red Army which prevailed over the German Army despite its gallant fighting spirit? Yes. Was it the highest German political and military command which brought about the defeat? Again, yes. But if at this early time a would-be analyst wanted to determine unequivocally which of the two was most instrumental in causing Germany's defeat, he would lack the necessary historical perspective as well as the requisite insight into a variety of events on both sides.

The recent war has reaffirmed only one fact: In any one-sided assault from the west, even the best of military forces will find it more than difficult to bring about the collapse of Russia.

U.S. GOVERNMENT PRINTING OFFICE : 1986 O - 155-677